STUDYING FOR BIOLOGY

STUDYING FOR BIOLOGY

ANTON E. LAWSON
Arizona State University

BRENDA D. SMITH, Series Editor
Georgia State University

HarperCollins*CollegePublishers*

Acquisitions Editor/Executive Editor: Ellen Schatz
Cover Designer: Ruttle Graphics, Inc./D. Setser
Electronic Production Manager: Angel Gonzalez Jr.
Publishing Services: Ruttle Graphics, Inc.
Electronic Page Makeup: Ruttle Graphics, Inc.
Printer and Binder: R. R. Donnelley & Sons Company
Cover Printer: The Lehigh Press, Inc.

Studying for Biology

Library of Congress Cataloging-in-Publication Data

Lawson, Anton E.
Studying for Biology / Anton E. Lawson
p. cm.
Includes index.
ISBN 0-06-500650-X
1. Biology —Study and teaching (Higher) 2. Thought and thinking.
I. Title.
QH51.L38 1995
574'.01—dc20 94-20073
CIP

96 97 9 8 7 6 5 4 3

ACKNOWLEDGMENTS

I would like to thank Brenda Smith for encouraging me to write this book and Marisa L'Heureux, Mark Paluch, and Jane Kinney for their helpful editorial suggestions. Of course the book would not have been possible without the many contributions of several of my former students. I gladly acknowledge my debt to them. A special thank you is also due to Jan Nagle, who typed the initial manuscript and designed several of the graphics.

Some of the material contained in this book is based in part on research supported by the National Science Foundation under Grant No. SED74-18950 and originally copyrighted by the Regents of the University of California. Any opinions, findings, and conclusions or recommendations expressed in this publication are those of the author and do not necessarily reflect the view of the National Science Foundation.

January 1995

A.E.L.

TO THE STUDENT

The purpose of this book is to help you become a successful student in college biology. To this end, it will help you gain insight into not only what biologists think, but how they think. Biologists are engaged in learning about the living world. Generally, they are quite good at learning, which is to say, they have learned how to learn. Therefore, this book will not only help you learn biology but, by providing insight into how biologists think, it will help you learn how to learn. The thinking/learning patterns of the biologist are not unique to biology. Consequently, study skills and the thinking/learning patterns you acquire will help you in other academic and nonacademic endeavors as well.

Chapter 1 discusses study skills for success in biology courses in terms of taking notes, studying, writing lab reports, and so on. The emphasis is on "deep processing" of information and reflective thought to increase understanding, retention, and transfer.

Chapter 2 discusses the nature of scientific thinking and its products, using familiar examples. The main message is that scientific thinking involves raising questions about causes and then attempting to answer these questions by generating (developing) and testing alternative hypotheses.

Chapter 3 focuses on the most important thinking patterns that biologists use to generate and test alternative hypotheses so that you can begin to acquire those thinking patterns. The examples will be less familiar than those in the first chapter, thus giving you a more general understanding of scientific thinking. This chapter also includes a self-test of scientific thinking. You will be able to compare your thinking patterns to those of several other students and to those of biologists.

Chapters 4 and 5 include a series of essays and problems that you can work through to develop competence in using the thinking patterns previously introduced.

Chapters 6–13 introduce the basic postulates of the major theories that have resulted from the use of the thinking patterns discussed in Chapters 2–5. These theories represent modern beliefs among informed biologists and provide the structure of present day biological thought. The major theories are presented in a brief and straight forward way. The benefit to you is that you can quickly gain an understanding of the important concepts and conceptual systems within biology. Another extremely important part of Chapters 6–13 is the inclusion of several examples of the use of the scientific thinking patterns introduced previously so that you can gain a still better understanding of their use.

Finally, you will find that most chapters include a list of terms that relate to biological phenomena. More than half of the chapters also include sample exam questions to show how biology concepts are translated into exam questions. The answers and explanations to end-of-chapter questions and the sample exam questions are found in the Appendix.

CONTENTS

CHAPTER 1 IMPROVING YOUR STUDY SKILLS 1

BECOMING AN ACTIVE AND REFLECTIVE LEARNER 2
MANAGING YOUR TIME 3
TAKING AND USING LECTURE NOTES 4
USING THE TEXTBOOK 7
PREPARING FOR AND TAKING EXAMS 12
MEMORIZING INFORMATION 21
WRITING A SCIENTIFIC PAPER 23
TIPS ON REVISING YOUR PAPER 25

CHAPTER 2 THE SCIENCE OF BIOLOGY 33

WHAT IS SCIENCE? 34
THE NATURE OF SCIENTIFIC THINKING: THE EARLY RISER 34
TRYING YOUR MIND AT DEDUCTIVE REASONING 38
THE NATURE OF HYPOTHESES 42
THE NATURE OF THEORIES 43

CHAPTER 3 DESCRIPTIVE AND HYPOTHETICAL THINKING PATTERNS 53

HOW GOOD ARE YOU AT SCIENTIFIC THINKING? 54
DESCRIPTIVE AND HYPOTHETICAL THINKING PATTERNS 60
HOMING IN SILVER SALMON 63
CREATIVE AND CRITICAL THINKING SKILLS 70

CHAPTER 4 IMPROVING YOUR SCIENTIFIC THINKING SKILLS: Part I 79

THE CASE OF CLEVER HANS 80
PROBABILISTIC THINKING 81
CORRELATIONAL THINKING 88

CHAPTER 5 IMPROVING YOUR SCIENTIFIC THINKING SKILLS: Part II 105

CHANCE AND CAUSAL RELATIONSHIPS 106
CONTROLLING VARIABLES 115
PROPORTIONAL RELATIONSHIPS 123

CHAPTER 6 THEORIES ABOUT THE NATURE OF LIFE 135

DO LIVING THINGS CONTAIN A "VITAL" FORCE? 136
A MODERN THEORY OF LIFE 139
A THEORY OF BIOLOGICAL ORGANIZATION AND EMERGENT PROPERTIES 141

CHAPTER 7 ORGANISM LEVEL THEORIES 151

A THEORY OF ORGANISM CLASSIFICATION 152
THEORIES OF ORGANISM BEHAVIOR 157
MENDEL'S THEORY OF THE INHERITANCE OF CHARACTERISTICS 158

CHAPTER 8 ORGAN AND SYSTEM LEVEL THEORIES 171

HARVEY'S THEORY OF BLOOD CIRCULATION 172
A THEORY OF HEART RATE REGULATION 173
A THEORY OF WATER RISE IN PLANTS 176

CHAPTER 9 CELL LEVEL THEORIES 183

CELL THEORY 184
A THEORY OF CELL REPLICATION 187
THEORIES OF SEXUAL REPRODUCTION AND MEIOSIS 191
TESTING HYPOTHESES ABOUT EMBRYOLOGICAL DEVELOPMENT 197
A MODERN THEORY OF EARLY EMBRYOLOGICAL DEVELOPMENT 199

CHAPTER 10 GENERAL MOLECULAR LEVEL THEORIES 209

KINETIC-MOLECULAR THEORY 210

GENERAL METABOLIC THEORY 216
A THEORY OF MOLECULAR MOVEMENT 222

CHAPTER 11 SPECIFIC MOLECULAR LEVEL THEORIES 235

A THEORY OF CELLULAR FERMENTATION AND RESPIRATION 236
A THEORY OF PHOTOSYNTHESIS 244
WATSON AND CRICK'S THEORY OF DNA STRUCTURE AND REPLICATION 251
A THEORY OF GENE FUNCTION (PROTEIN SYNTHESIS) 256

CHAPTER 12 ECOLOGICAL LEVEL THEORIES 271

A THEORY OF POPULATION GROWTH AND CRASH 272
THEORIES OF POPULATION REGULATION 276
A THEORY OF ECOSYSTEM DYNAMICS 281
A THEORY OF SUCCESSION 286

CHAPTER 13 THEORIES OF ORGANIC CHANGE 305

A THEORY OF THE ORIGIN OF LIFE ON EARTH 306
ORGANIC EVOLUTION THEORY 309
DARWIN'S THEORY OF NATURAL SELECTION AND A SYNTHETIC THEORY OF EVOLUTION 313

GLOSSARY 327

APPENDIX: ANSWERS TO QUESTIONS AND SAMPLE EXAM QUESTIONS 345

INDEX 386

CHAPTER 1

IMPROVING YOUR STUDY SKILLS

GETTING FOCUSED

- *Become an active and reflective learner.*
- *Manage your time.*
- *Take and use lecture notes.*
- *Use the textbook.*
- *Prepare for and take exams.*
- *Write a scientific paper.*

Earning an A in biology requires three things. First, you need to know how to study. That is the subject of Chapter 1. Second, you need to know how biologists think. That is the subject of Chapters 2–5. And third, you need to know the major conceptual systems, called theories, that are central to what biologists think. You will find those in Chapters 6–13.

Contrary to what some people may have led you to believe, you do not need to have an innately high IQ to be successful. If you know how to study, know how and what biologists think, and are sufficiently motivated to put forth a modest amount of effort, you will earn that A. Happily, there is a bonus. It turns out that the studying and thinking skills and the conceptual knowledge of the biologist are useful in all walks of life where questions arise and ideas should be tested. Therefore, if you can improve these three elements of learning, you will be more successful in other courses, indeed in life in general.

Have you ever taken a course in which you read and reread the text, attended all lectures and took endless notes, carefully completed all assignments and conscientiously studied for exams, but still earned a poor grade? If so, you may need to rethink your approach to studying. This chapter presents a series of strategies for using your time effectively so that you really learn the material and earn a good grade. It will first provide suggestions on how you can become a more active and reflective learner and better manage your time. You will also find specific suggestions for taking lecture notes, using the textbook, preparing for and taking exams, improving your memory, and writing a good scientific paper.

BECOMING AN ACTIVE AND REFLECTIVE LEARNER

If you are going to learn how to swim, you must first get into the water. The same holds true in learning biology. Standing on the shore will not cut it. You have to dive in. Here are some specific suggestions:

- Sit in the front of the lecture room and laboratory. When possible, ask questions and participate in discussions.
- Try to figure out the instructor's purpose for each assignment. If uncertain, ask the instructor.
- While reading and studying, try to sort the important from the unimportant information.

- As you read your text and notes, ask and answer questions and try to predict what questions are likely to be on exams.
- Try to make connections among the lectures, the laboratories, and the text.
- Review returned assignments and exams to discover and correct mistakes.
- React to, evaluate, and attempt to apply what you have learned to other courses and to everyday life.
- Form a study group with other students to share experiences, ideas, and questions.
- Use spare time, such as when you are going to and from classes, are about to fall asleep at night, or just after you have awakened in the morning, to reflect on what you are learning and its implications.

Being reflective just might be the most important element for success. Students who do poorly seldom think about what they are learning. Good students turn off the music and the television and they talk to themselves. They ask themselves questions about what they are learning, and talk to themselves to try to get answers.

Answers may not come right away, but the mind has the amazing ability to come up with answers even when a person is not consciously thinking. In fact, for really difficult questions, it may be necessary to allow the subconscious mind to take over and consider the question before an answer pops up. Perhaps this is why so many of the great ideas in science have come to scientists when they were engaged in decidedly unscientific pursuits, such as taking a bath or sleeping!

MANAGING YOUR TIME

Your time is valuable, so use it wisely. Here are some guidelines:

- Get a calendar and construct a daily and weekly schedule for the entire semester.
- Get a loose-leaf notebook to record lecture, laboratory, and textbook notes, and to collect returned assignments and exams.
- Keep a list of things to do and revise it often.

- Find a regular place to study free from distractions (e.g., music, TV, other people). Train others not to disturb you; or study where people cannot find you.
- Schedule regular study sessions in blocks of 40–50 minutes with 5–10 minute breaks.
- Vary your study activities often enough so that you do not get bored and lose your ability to concentrate.
- Make sure you schedule enough study time. One rule of thumb calls for two hours of study for every hour of class, but you can increase or decrease the time depending on specific course demands.
- Plan to attend all classes and take good notes.
- Schedule time to review and edit lecture notes as soon as possible after each lecture.
- Ask for help from classmates and/or the instructor as soon as problems arise so that you do not fall behind, particularly if your performance on the first exam is not satisfactory.
- Schedule time to review your notes several times each week.
- Get ample sleep, take time to exercise, to relax and just goof off, and remember to eat a nutritious and balanced diet.

TAKING AND USING LECTURE NOTES

In most biology courses, much of what you are tested on comes from the lectures and laboratories. Therefore, above all else, attend every lecture and laboratory and take good notes. It may be possible to miss a lecture and get notes from someone else, but no matter how good that person's notes are, they cannot substitute for yours. Here is how to take good notes:

- Use a large loose-leaf notebook so that you have plenty of room for revising and adding information to your notes. The loose-leaf feature allows you to insert course hand-outs and exams in the proper order.
- Use ink and write legibly so that you have a permanent and readable set of notes.

- Date and title the notes from each lecture.
- Write your notes on the right-hand pages only. Use the left-hand pages when you review and edit your notes.
- Write down as much of what the instructor says as possible, but focus on ideas/concepts and not on facts and details. Underline key terms and points stressed by the instructor.
- Use an outline format with indentations and main topics above main ideas, above details, and so forth, but do not be concerned when some elements do not fit the outline pattern. Just write them down and worry about organization later. You can use roman numerals (I, II, III), capital letters (A, B, C), and Arabic numerals (1, 2, 3), but they are not essential.
- Leave some blank spaces to help with readability and to allow yourself space for filling in details later.
- Use abbreviations and incomplete sentences when possible.
- Do not tape record lectures to avoid taking notes. This only delays the need to take notes and wastes time.
- Do not recopy your notes. This wastes time that could be better spent reviewing.
- Review and edit your notes within an hour or so after the lecture. You will be amazed to discover that you can recall a considerable amount of detail from the lecture that, at the time, you were unable to get down. Use the review time to record those details you think are important. The review will also allow you to identify points that are unclear, so that you can figure them out before the next lecture.
- When you identify points that are unclear in your notes, you have four ways to reach understanding. First, try to think of a question that focuses on what is unclear. Then write the question down on a sheet of paper and try answering it yourself in writing. Writing down your thoughts and then going back and reading and thinking about what you wrote will help you to "think about your thinking." This is the key to being a reflective learner, filling in gaps, making connections, resolving contradictions, and seeing implications. Second, go to your textbook to see if it helps and/or find one or more classmates with whom to discuss the material. Third, leave the question and go on to something

else. Your subconscious mind may take over and answer it for you. A good time to raise the question is just before falling asleep at night. This gives the subconscious mind all night to think about possible answers. By morning you may wake up with the answer. Finally, do not hesitate to take any remaining question(s) to your instructor.

- After you have edited your notes, use the left margin to write a few key words or phrases that summarize or label key ideas. You can use these key words or phrases later to help you study your notes. After reading the notes several times, cover them up with a sheet of paper, leaving only the key words or phrases visible. Then use these cues to see if they can trigger recall of the information you have covered up. The key words and phrases can also be used to prepare summary sheets from several lectures. The summary sheets can then be used for a quick review.
- Review your notes often. Review the notes from lecture one before attending lecture two. Review the notes from lectures one and two before attending lecture three. Review the notes from lectures one, two, and three before attending lecture four, and so on, until you are tested on the material. This may sound like a lot of work, but it makes recall and synthesis easier and practically eliminates the need to spend extra time studying before exams.

The act of writing about what you are trying to learn can be very helpful. Read what a student named Tim had to say about his experience in using writing to make difficult ideas easier:

The writing enabled me to know what I knew and to figure out the difficult ideas on my own. I realized I could listen to myself think while I wrote my ideas on paper. When I wrote, I also felt as though I let the ideas at the forefront of my brain out onto the paper so that the ideas in the back of my brain could move up into the front. Once those first ideas, the conscious ones, were out of my head, the ideas in the back of my brain, the subconscious ones, had room to move to the forefront to become the conscious ideas. And when I wrote these out, more ideas moved into the forefront. The new ideas had been too far in the back of my brain, and I couldn't get to them until I made room. With all those new ideas now available to me, I felt as if I were discovering new things which I didn't know I knew. At the same time, I was making connections between one *bit of information and another, sorting and reshuffling and discovering even more. When I transferred the idea to paper, I could actually* see *what I knew about the idea. I found that writing eliminated my need to study the night before a test. (pp. 33–35)**

You, too, may find writing to be an effective strategy, particularly if you are not familiar with the hypothetical thinking patterns discussed in the following chapters. Becoming a reflective thinker requires that you "think about your thinking." Writing is a good way to do this.

USING THE TEXTBOOK

In addition to lectures and laboratories, your biology course will probably have a textbook. Textbooks are generally, but not always, of secondary importance. If your course has a text, you should ask your instructor to indicate how he or she intends for it to be used. In most cases, it will be used to supplement and reinforce points the instructor makes in the lectures and laboratories.

Most introductory biology textbooks are organized in one of two ways. One organization starts by discussing very small and relatively simple things, such as atoms and molecules, and then goes on to discuss progressively larger, more complex things, such as cells, organs, organisms, and ecosystems. The second starts with the big things and goes in reverse order to the small. The first approach is sometimes called, appropriately enough, the small-to-large, or micro-to-macro, approach, while the second is called the large-to-small, or macro-to-micro, approach.

Both approaches recognize a very important pattern, which is that biological phenomena are organized into distinct levels of size and complexity. Starting from the small end of things, we find that matter consists of atoms that may combine with other atoms to form molecules, which in turn may combine to form organelles, which in turn may combine to form cells, and so on. The levels, going from small to large, are as follows: atoms → molecules → organelles → cells → tissues → organs → organ systems → organisms → populations → biological communities → ecosystems → biomes → biosphere. (More details can be found in Chapter 6.) Each of these levels of organization presents its own distinct phenomena, which raise their own distinct

*A. M. Wotring and R. Tierney, *Two Studies of Writing in High School Science* (Berkeley, CA: Bay Area Writing Project, Regents of University of California at Berkeley, 1981).

sets of questions. Recognizing that the universe consists of these levels of complexity is extremely important, because it will enable you to organize vastly different sets of phenomena in one conceptual system. In other words, it keeps you from losing the "forest for the trees" when reading the text, and will help you make connections with topics from other levels.

- Therefore, the first thing you should do is survey your text to determine its overall organization in terms of these levels. Does the text take the small-to-large approach, does it take the large-to-small approach, or does it take some combination of the two? Figures 1.1 to 1.3 show the contents of three popular introductory texts. Notice that in the margins I have identified the principle level(s) of organization discussed in each chapter. The first text, by Audesirk & Audesirk, in Figure 1.1, takes the micro-to-macro approach, while the second text, by Levine and Miller, in Figure 1.2, takes more of a macro-to-micro approach. The third text, by Mix, Farber, and King, shown in Figure 1.3, starts at the organism level, then goes from macro to micro in Chapters 5–21, then it goes to the population level and organism levels for Chapters 22–28. Finally, Chapters 29–40 are largely devoted to the organ and organ system levels.

- Before you read a chapter, survey it to identify its major topics. To do this, first read the chapter title to find out the level(s) of organization discussed. For example, a glance through a chapter titled "Life's Origins and Microevolutionary Trends" may reveal that it is mainly about the population level of organization, with discussions of the origin of life that center on molecular and cellular level phenomena. Before you read the chapter, turn the title into a question or questions. For this chapter, you can raise these two questions: How did life originate? What are some microevolutionary trends? When you read the chapter, your goal becomes one of answering these and related questions.

- After you have read the chapter title, determined the level(s) of organization involved, and thought of one or two major questions, read the chapter summary to find out what major new terms are introduced. Then go back to the start and read the section headings, paying attention to pictures, tables, graphs, and their captions. The main intent here is to get the lay of the land. What is the chapter about? How are the topics going to be presented?

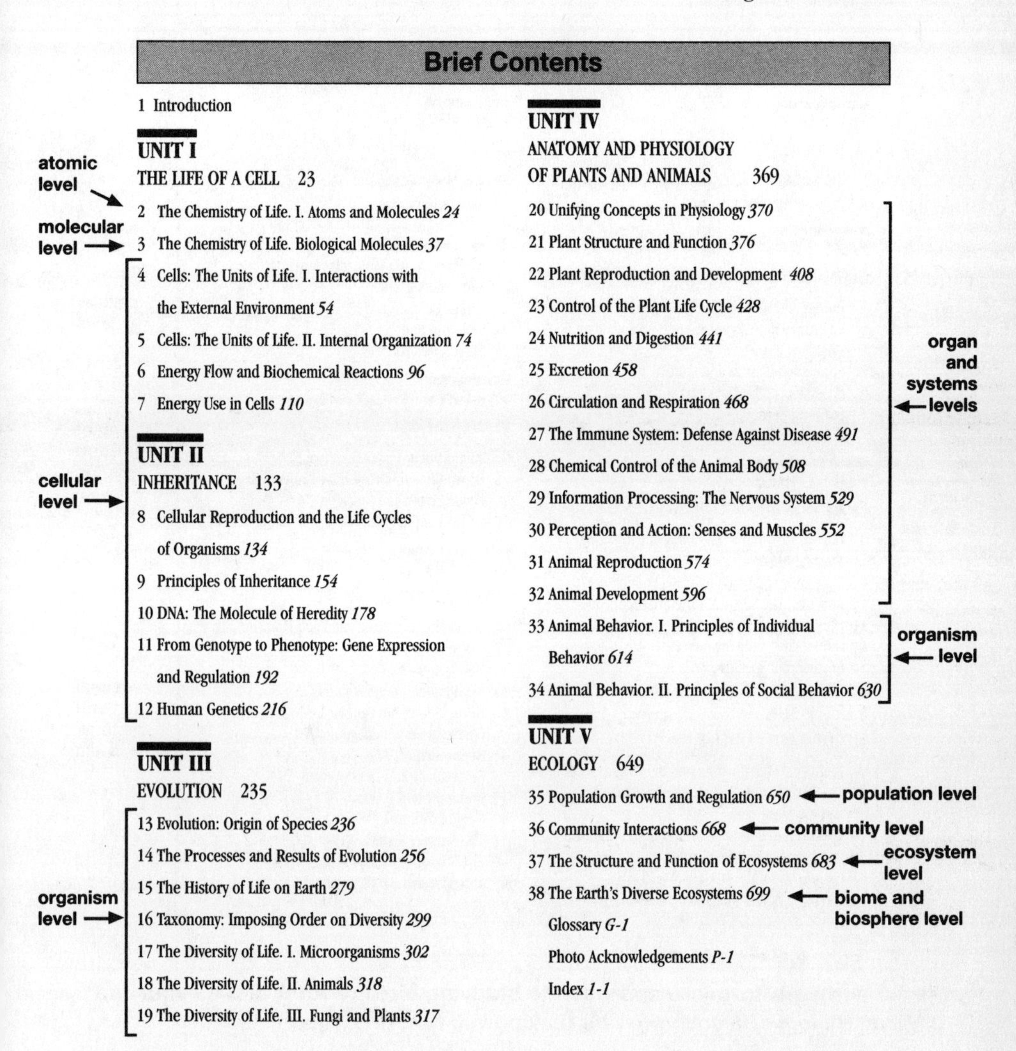

Brief Contents

1 Introduction

UNIT I
THE LIFE OF A CELL 23
2 The Chemistry of Life. I. Atoms and Molecules *24*
3 The Chemistry of Life. Biological Molecules *37*
4 Cells: The Units of Life. I. Interactions with the External Environment *54*
5 Cells: The Units of Life. II. Internal Organization *74*
6 Energy Flow and Biochemical Reactions *96*
7 Energy Use in Cells *110*

UNIT II
INHERITANCE 133
8 Cellular Reproduction and the Life Cycles of Organisms *134*
9 Principles of Inheritance *154*
10 DNA: The Molecule of Heredity *178*
11 From Genotype to Phenotype: Gene Expression and Regulation *192*
12 Human Genetics *216*

UNIT III
EVOLUTION 235
13 Evolution: Origin of Species *236*
14 The Processes and Results of Evolution *256*
15 The History of Life on Earth *279*
16 Taxonomy: Imposing Order on Diversity *299*
17 The Diversity of Life. I. Microorganisms *302*
18 The Diversity of Life. II. Animals *318*
19 The Diversity of Life. III. Fungi and Plants *317*

UNIT IV
ANATOMY AND PHYSIOLOGY OF PLANTS AND ANIMALS 369
20 Unifying Concepts in Physiology *370*
21 Plant Structure and Function *376*
22 Plant Reproduction and Development *408*
23 Control of the Plant Life Cycle *428*
24 Nutrition and Digestion *441*
25 Excretion *458*
26 Circulation and Respiration *468*
27 The Immune System: Defense Against Disease *491*
28 Chemical Control of the Animal Body *508*
29 Information Processing: The Nervous System *529*
30 Perception and Action: Senses and Muscles *552*
31 Animal Reproduction *574*
32 Animal Development *596*
33 Animal Behavior. I. Principles of Individual Behavior *614*
34 Animal Behavior. II. Principles of Social Behavior *630*

UNIT V
ECOLOGY 649
35 Population Growth and Regulation *650*
36 Community Interactions *668*
37 The Structure and Function of Ecosystems *683*
38 The Earth's Diverse Ecosystems *699*
Glossary *G-1*
Photo Acknowledgements *P-1*
Index *I-1*

Figure 1.1 A micro-to-macro approach to studying biology. After G. Audesirk and T. Audesirk, *Biology: Life on Earth,* 2nd. ed. (New York: Macmillan, 1989).

- After you have surveyed the chapter, begin reading section by section. Before you read a section, turn the section heading into a question so that, once again, you are reading with a distinct purpose, which is of course to answer the question. For example, for a section headed "Circadian Rhythms," ask, What are circadian rhythms? Although this may seem simple, it works. When

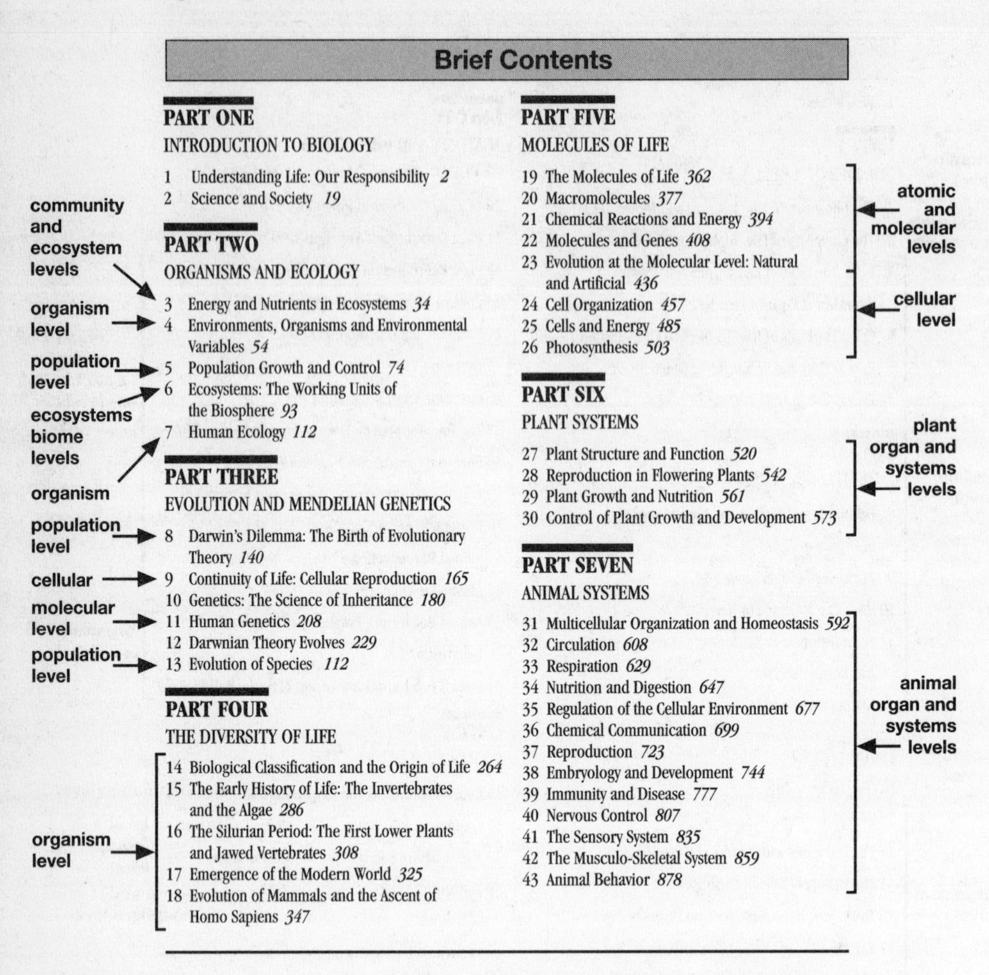

Brief Contents

PART ONE
INTRODUCTION TO BIOLOGY

1 Understanding Life: Our Responsibility *2*
2 Science and Society *19*

PART TWO
ORGANISMS AND ECOLOGY

3 Energy and Nutrients in Ecosystems *34*
4 Environments, Organisms and Environmental Variables *54*
5 Population Growth and Control *74*
6 Ecosystems: The Working Units of the Biosphere *93*
7 Human Ecology *112*

PART THREE
EVOLUTION AND MENDELIAN GENETICS

8 Darwin's Dilemma: The Birth of Evolutionary Theory *140*
9 Continuity of Life: Cellular Reproduction *165*
10 Genetics: The Science of Inheritance *180*
11 Human Genetics *208*
12 Darwnian Theory Evolves *229*
13 Evolution of Species *112*

PART FOUR
THE DIVERSITY OF LIFE

14 Biological Classification and the Origin of Life *264*
15 The Early History of Life: The Invertebrates and the Algae *286*
16 The Silurian Period: The First Lower Plants and Jawed Vertebrates *308*
17 Emergence of the Modern World *325*
18 Evolution of Mammals and the Ascent of Homo Sapiens *347*

PART FIVE
MOLECULES OF LIFE

19 The Molecules of Life *362*
20 Macromolecules *377*
21 Chemical Reactions and Energy *394*
22 Molecules and Genes *408*
23 Evolution at the Molecular Level: Natural and Artificial *436*
24 Cell Organization *457*
25 Cells and Energy *485*
26 Photosynthesis *503*

PART SIX
PLANT SYSTEMS

27 Plant Structure and Function *520*
28 Reproduction in Flowering Plants *542*
29 Plant Growth and Nutrition *561*
30 Control of Plant Growth and Development *573*

PART SEVEN
ANIMAL SYSTEMS

31 Multicellular Organization and Homeostasis *592*
32 Circulation *608*
33 Respiration *629*
34 Nutrition and Digestion *647*
35 Regulation of the Cellular Environment *677*
36 Chemical Communication *699*
37 Reproduction *723*
38 Embryology and Development *744*
39 Immunity and Disease *777*
40 Nervous Control *807*
41 The Sensory System *835*
42 The Musculo-Skeletal System *859*
43 Animal Behavior *878*

Figure 1.2 A macro-to-micro approach to studying biology. After J. S. Levine and K. R. Miller, *Biology: Discovering Life* (Lexington, MA: D.C. Heath, 1991).

you are finished reading the section, ask yourself your question. If you cannot answer the question, you will know right away that you did not comprehend the section and need to reread it before going on. If, after rereading, you still are not comprehending, try one or more of these steps: Go back a few sections to try to pick up the author's ideas. Read a few paragraphs ahead to see what is next. Use the glossary or a dictionary to look up any words that you do not know. Reread the difficult sentences or paragraphs aloud with expression to increase your level of concentration. And finally, if you still do not understand, try to

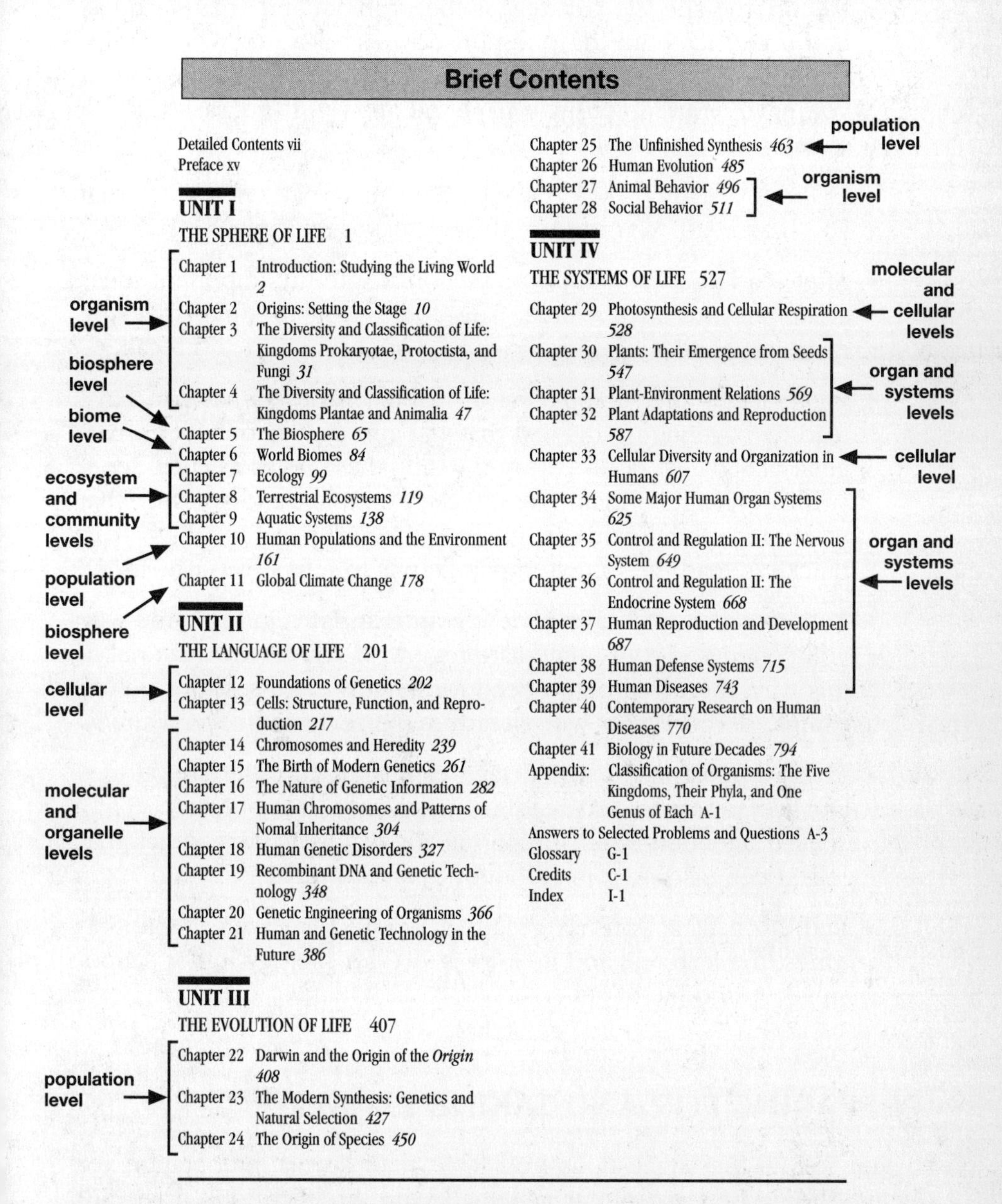

Brief Contents

Detailed Contents vii
Preface xv

UNIT I
THE SPHERE OF LIFE 1

Chapter 1 Introduction: Studying the Living World *2*
Chapter 2 Origins: Setting the Stage *10*
Chapter 3 The Diversity and Classification of Life: Kingdoms Prokaryotae, Protoctista, and Fungi *31*
Chapter 4 The Diversity and Classification of Life: Kingdoms Plantae and Animalia *47*
Chapter 5 The Biosphere *65*
Chapter 6 World Biomes *84*
Chapter 7 Ecology *99*
Chapter 8 Terrestrial Ecosystems *119*
Chapter 9 Aquatic Systems *138*
Chapter 10 Human Populations and the Environment *161*
Chapter 11 Global Climate Change *178*

UNIT II
THE LANGUAGE OF LIFE 201

Chapter 12 Foundations of Genetics *202*
Chapter 13 Cells: Structure, Function, and Reproduction *217*
Chapter 14 Chromosomes and Heredity *239*
Chapter 15 The Birth of Modern Genetics *261*
Chapter 16 The Nature of Genetic Information *282*
Chapter 17 Human Chromosomes and Patterns of Nomal Inheritance *304*
Chapter 18 Human Genetic Disorders *327*
Chapter 19 Recombinant DNA and Genetic Technology *348*
Chapter 20 Genetic Engineering of Organisms *366*
Chapter 21 Human and Genetic Technology in the Future *386*

UNIT III
THE EVOLUTION OF LIFE 407

Chapter 22 Darwin and the Origin of the *Origin* *408*
Chapter 23 The Modern Synthesis: Genetics and Natural Selection *427*
Chapter 24 The Origin of Species *450*
Chapter 25 The Unfinished Synthesis *463*
Chapter 26 Human Evolution *485*
Chapter 27 Animal Behavior *496*
Chapter 28 Social Behavior *511*

UNIT IV
THE SYSTEMS OF LIFE 527

Chapter 29 Photosynthesis and Cellular Respiration *528*
Chapter 30 Plants: Their Emergence from Seeds *547*
Chapter 31 Plant-Environment Relations *569*
Chapter 32 Plant Adaptations and Reproduction *587*
Chapter 33 Cellular Diversity and Organization in Humans *607*
Chapter 34 Some Major Human Organ Systems *625*
Chapter 35 Control and Regulation II: The Nervous System *649*
Chapter 36 Control and Regulation II: The Endocrine System *668*
Chapter 37 Human Reproduction and Development *687*
Chapter 38 Human Defense Systems *715*
Chapter 39 Human Diseases *743*
Chapter 40 Contemporary Research on Human Diseases *770*
Chapter 41 Biology in Future Decades *794*
Appendix: Classification of Organisms: The Five Kingdoms, Their Phyla, and One Genus of Each A-1
Answers to Selected Problems and Questions A-3
Glossary G-1
Credits C-1
Index I-1

Figure 1.3 A varied approach to studying biology. After M. C. Mix, P. Farber, and K. I. King, *Biology: The Network of Life* (New York: HarperCollins, 1992).

identify the specific part that is the problem and create a question about it. Write the question down and try to write an answer. Once you have written something and reflected on what you have written, you may have the answer. If not, ask a classmate or the instructor.

- In the text margins or on a separate sheet, make notes on what you have read. You can use these later for review.
- After reading a paragraph or section, recite the important ideas aloud in your own words. This forces you to summarize the main ideas. Reciting aloud increases concentration and recall.
- Immediately after completing a chapter, review what you have read by covering up the text material and leaving only your marginal or separate notes exposed. Using these notes as cues, see if you can recall the main ideas, facts, examples, and so forth, that were presented. If you can, go on to the next page. If not, you can quickly reread the forgotten paragraphs to refresh your memory.
- You should periodically review your text notes in the same way you conducted your immediate review. If you review material in this way, there will be no need for lengthy cram sessions before exams, because you will already have reviewed several times.
- Because most instructors design exams around the laboratories and lectures, with text readings used only as supplements, pay special attention when preparing for exams to text topics that have been discussed in laboratories and lectures.
- Reflect on what you have read, its connections with the laboratories, the lectures, and its importance for biology, other school subjects, and life in general.

PREPARING FOR AND TAKING EXAMS

If you follow the suggestions you have read so far, preparation for exams should be easy. As far as your lecture and textbook notes are concerned, one brief review should be enough. In addition to one last brief review of your notes, the best thing you can do to prepare for exams is to locate one or more of the instructor's exams from previous semesters. These exams will provide you with insight into the format of the exam and what the instructor thinks is important. Remember, the most important thing for you to do is to try to get into the instructor's mind, look at things from his or her perspective, and try to predict just what questions will be on the exam. In addition to looking at past exams, ask the instructor to tell you as much about the exam as he or she is willing. Is the format multiple choice, true/false, essay, or short answer. How many questions will be on the exam? What does he or she think are the most important concepts, topics, and issues? Comparing ideas about what to expect with other students can also be very helpful.

Levels of Exam Questions

In addition to these issues, there remains the issue of difficulty levels of exam questions. How tough will the exam be? Should I memorize or do I really have to "know" the material?

Just what makes exam questions easy or difficult? To find out, let us consider several different types of exam questions. Exam questions are generally classified into the following types or levels of difficulty because they reflect use of increasingly complex thinking skills.

1. **Knowledge.** Exam questions written at the knowledge level require students to recall only specific bits of information.

2. **Comprehension.** Comprehension questions require an understanding of what is being communicated, but students do not need to relate it to other knowledge or discuss its implications.

3. **Application.** These questions require students to apply concepts, procedures, and/or thinking skills in specific situations.

4. **Analysis.** Analysis questions require students to break down a statement, statements, or data set into parts so that they make clear a hierarchy of ideas and/or show the relationships among ideas; students must also recognize unstated assumptions, distinguish facts from hypotheses, and check the consistency of hypotheses with given information.

5. **Synthesis.** Synthesis questions require students to put together parts to form a whole, arrange and combine the parts to reveal a pattern or structure not present in the individual parts, formulate appropriate hypotheses based upon an analysis of factors involved, and modify such hypotheses in light of new considerations.

6. **Evaluation.** Here students must make judgments about the value of given materials and/or methods for specific purposes, compare major theories, and find logical fallacies in arguments.

Now read the following six sample exam questions and try classifying each as one of these levels of difficulty. Record your classifications in the space provided following question 6. You do not have to answer the questions themselves.

Question 1 During the 1920s, a population of spotted crabs was known to inhabit the white sandy beaches near a volcano on one of the Hawaiian Islands. The spotted crabs were observed feeding off plants deposited on the beaches by ocean waves. Occasionally, seagulls were observed capturing and eating some of the crabs. When first observed, about 90% of the crabs were nearly completely white, with only a few small black spots on their claws. About 8% of them were white with many black spots, while 2% of the crabs had so many black spots they appeared almost completely black.

In 1930 the volcano erupted, sending a lava flow across the beach out into the water. The lava cooled and blocked the ocean currents that had deposited the white sands on the beach. Black sand from other currents began to accumulate on the beach, until within a few years it was completely covered with black sand.

By the 1950s, nearly 95% of the spotted crab population was composed of crabs that were completely black. About 4% of the crabs were white with many black spots, while 1% of the crabs were white with only a few black spots.

Use the theory of natural selection to explain the change in the most frequent color of the spotted crabs from white to black. Be sure to discuss the role of biotic potential, variation, limiting factors, and heredity.

Question 2 The name given to the upper part of the human small intestine is

a. colon.

b. duodenum.

c. villus.

d. cecum.

Question 3 Fifty pieces of various parts of plants were placed in each of five sealed containers of equal size. At the start of the experiment, each jar contained 250 units of CO_2. The amount of CO_2 in each jar at the end of two days is shown in Table 1.1.

On the basis of the data in the table, a fair test of the amount of CO_2 used per day at two different temperatures could be made by comparing which jars? Explain why you chose those jars.

TABLE 1.1 CO_2 DATA

JAR	PLANT	PLANT PART	LIGHT COLOR	TEMPERATURE °C	CO_2 REMAINING
1	willow	leaf	blue	10	200
2	maple	leaf	purple	23	50
3	willow	root	red	18	300
4	maple	stem	red	23	400
5	willow	leaf	blue	23	150

Question 4 An animal that has skin that is not permeable to salt goes for a swim in the Pacific Ocean and gains weight. What would happen to that animal if it went for a swim in Lake Tahoe (a freshwater lake)? It would probably

a. lose weight.

b. gain weight.

c. neither gain nor lose weight.

d. not enough information to tell.

Question 5 Experiments such as the one described by Stanley Miller indicate that no special supernatural processes or powers are necessary to get simple inorganic molecules to combine into larger, more complex, organic molecules. Apparently, all that is needed is for a number of these small inorganic molecules to accumulate in one place and that energy be supplied. This starts molecular collisions until the small molecules "spontaneously" combine. In a sense, then, this theory of the origin of life involves the idea of spontaneous generation. But the experiments of Redi, Spallanzani, and Pasteur argued that the theory of spontaneous generation is wrong. Instead, the theory of biogenesis is correct. Are the theory of biogenesis and the present theory of the origin of life contradictory to each other or not? Explain.

Question 6 Which of the following is an example of a primary consumer?

a. mold that grows on bread

b. hawks that eat mice

c. mice that eat leaves

d. bacteria that eat dead logs

Record your classification of the six exam questions here.

Question 1 is a ______ level question.
Question 2 is a ______ level question.
Question 3 is a ______ level question.
Question 4 is a ______ level question.
Question 5 is a ______ level question.
Question 6 is a ______ level question.

Now compare your classifications with those that follow.

In my opinion, Question 1 is a synthesis level question because it requires students to synthesize (put together) a number of concepts (e.g., natural selection, biotic potential, variation) to explain a specific phenomenon. The lower levels of knowledge, comprehension, application, and analysis are also involved in the synthesis of the concepts to explain the color change satisfactorily.

Question 2 is a knowledge level question. No comprehension is required. All students need to do is recall the correct label for the particular structure.

Question 3 is an analysis level question because it requires students to apply the method of identifying and controlling variables and combinatorial thinking (which you will learn about in Chapters 3–5) to analyze the data correctly and conclude that jars 1 and 5 should be used. Jars 1 and 5 should be used because they are identical except for the temperature (i.e., same type of plant, same plant part, and same color of light).

Question 4 requires students to apply the concept of osmosis to conclude correctly that the animal will gain weight. (The animal must have a lot of salt in its cells because the water from the salty ocean entered its cells, so the freshwater would enter even faster.) Therefore, question 4 should be classified at the application level.

I classified Question 5 as an evaluation level question because students are required to evaluate the theory of biogenesis and the theory of the origin of life in light of the experimental evidence gathered by Redi, Spallanzani, and Pasteur.

And lastly, I classified Question 6 at the comprehension level because it requires students to comprehend the concept of primary consumer in order to select the correct answer; that is, a primary consumer is any animal that feeds directly off plants. Mice are animals and leaves are plants: therefore, choice c is correct.

If you did not classify the exam questions in precisely the same way I did, do not be concerned. The point of this section is not to make you an expert at classifying exam items. Rather, it is that instructors differ widely in the difficulty level of questions they typically ask. Because different level questions require different thinking skills and different preparation, it becomes extremely important for you to determine, as soon as possible, what level of questions you are most likely to be asked. If the majority of questions are low-level knowledge and comprehensive questions, you will need to work hard to memorize lots of separate bits of information. If, on the other hand, the majority of questions are at the higher levels and require higher-order thinking skills, then you will need to focus on major concepts and spend considerable time reflecting on their connections and implications. Of course, this is where the real payoff of an educational experience lies. Therefore, we can hope that your instructor writes exam questions that require use of higher-order thinking skills. Although you will soon forget isolated bits of information, learning to use higher-order thinking skills will help you in nearly all future pursuits.

With these general remarks in mind, here are some specific suggestions on how to take exams:

- Arrive at the exam room early enough to get a seat near the front, but not so early that you become anxious because of a long wait. Sitting near the front will ensure that you receive an exam quickly. It also makes it easier to hear last-minute instructions and will help you avoid distractions from other students who arrive late or leave early.
- Before you start, listen to any last-minute verbal instructions and read written instructions very carefully. If you are unclear about

what you are supposed to do, make sure to raise your hand high and ask. Above all, do not proceed until the directions are clear. This same rule applies to individual exam questions. When in doubt, ask for clarification and pay careful attention to the instructor's reply. Often the reply will provide valuable clues.

- Look over the entire exam before you start and make a judgment as to how best to spend your time. Plan to answer the easy and/or brief questions first. This helps build confidence and cures anxiety so that you are able to tackle the more difficult questions successfully.

- If you come to a question that you cannot answer, skip it and come back to it later. But do not forget to get back to it! Do not assume that the instructor is trying to trick you. If you do, you may read too much into the question. The idea is to try to understand the point behind the question.

- If you finish and have time, review the entire exam to make sure that you have not skipped any questions or answered carelessly or incorrectly. Do not hesitate to add points of clarification or change answers that you think are incorrect. But be careful not to change an answer just because you are unsure of your original response. If you think your first response was wrong, change it. If you do not know, leave it alone.

- You should use the exam as a learning tool. This can be done in two ways. First, immediately after the exam, while the questions are still fresh in your mind, discuss them with several of your classmates. The discussion will help you compare your interpretations of questions and your lines of reasoning and answers with others. Second, when the exam is passed back, read the questions you missed. Were your mistakes due merely to carelessness? If so, slow down next time and try to be more careful. Did you not understand some ideas? If so, go back to your notes, your text, your classmates, or the instructor to clarify the misunderstanding. If you did poorly on essay responses, try to find some responses that were graded better than yours and compare them. What ideas, patterns of organization and/or presentation were better than yours? Keep in mind that most instructors give more weight to later exam scores; therefore, if you can correct your mistakes and improve during the semester, your final grade may be based upon your most recent exam grade, rather than on an average of all exam grades.

In addition to these six general test-taking strategies, here are some strategies for specific types of test items.

Strategies for True/False Questions

- Because the majority of statements about biological phenomena have exceptions, most statements on true/false exams that contain such words as *always, never, all,* or *none* are false.
- Any statement that contains unfamiliar terms or facts is probably false. If you have studied the material carefully, you should at least be familiar with the terms or facts that have been presented; therefore, if you have never heard of it, the instructor probably made it up as a false distractor.
- If the statement contains two parts, and one part is true and the other part false, the entire statement should be marked false.
- If you simply cannot decide whether a statement is true or false, it is better to guess true because it is difficult for an instructor to write believable false statements.

Strategies for Matching Questions

- First, quickly read both columns to familiarize yourself with all the items to be matched.
- Start by matching items that you are sure of, but make sure that you read all of the possibilities before selecting a match.
- After you have selected an item, cross it off so that you do not consider it further.
- Continue by matching items that you are sure of, leaving the questionable ones for last. The fewer questionable items you leave yourself with, the greater are your chances of guessing the remaining items correctly.

Strategies for Multiple-Choice Questions

- Read all the answer choices before selecting one. Keep in mind that most instructors expect you to pick the best choice even if they do not say so. Therefore, if you do not read all of the choices, you may miss the best one.

- Use the process of elimination. To do this, first cross off the answer choices that you are sure are wrong. With luck or correct reasoning, you will arrive at a single correct choice. If not, you have at least increased your chances of a correct guess by reducing the number of possibilities.
- Some multiple-choice questions include answer choices that are combinations of previous choices (e.g., a and b, all of the above, none of the above). When answering such questions, treat each of the initial choices as true/false statements and mark each true or false before going to the next. If, for example, all the initial choices are marked true, then you should answer "all of the above."
- As with true/false questions, look for key qualifying words. Choices with extreme words such as *all, none, always, never, entirely,* and *completely* tend to be false. Choices with more moderate words such as *often, seldom, probably, more,* and *fewer* tend to be true.
- Again as with true/false questions, choices that contain unfamiliar terms or facts are probably false and, therefore should not normally be selected.
- Longer answer choices are more often correct than shorter ones. This is because instructors generally try to word the correct choice carefully and completely. Therefore, if you are reduced to guessing, select the longest choice that seems reasonable.

Strategies for Essay Questions

- Before you answer an essay question, read the question very carefully word by word, taking notes on ideas, arguments, and examples that you will want to include in your response.
- Use your notes to construct a brief outline of your response. This is the time to reflect on what you know and carefully organize that knowledge in the best way possible before you start writing.
- Use a pen and write as neatly and quickly as possible on only the right-hand page.
- Use standard paragraph form, which contains one central idea per paragraph with supporting arguments and/or examples. Be

as specific as possible with your examples and use proper citations when possible.

- Make your key ideas easy to find by using headings, underlines, and/or blank spaces.
- Be careful that you do not get carried away with what you are writing and drift off the issue(s). What you write may be entirely correct and well written, but if it is not to the point, it is not likely to get you a good grade.
- Proofread your answer to correct illegible words, misspellings, incorrect punctuation, awkward sentences, and poor or missing transitions. Unclear or distracting writing reduces your ability to communicate your knowledge to the instructor. Although your ideas, arguments, evidence, and your organizational skills are of utmost importance, it is a rare instructor who does not also, either consciously or subconsciously, take these other elements of effective writing into consideration.

MEMORIZING INFORMATION

Of course, one would hope that your biology course will require little in the way of rote memorization (i.e., acquiring disconnected information in a mechanical way with little or no thought). Most instructors realize that stocking your memory with isolated bits of information does little or no good, except perhaps to make you better at such things as winning games of Trivial Pursuit. Nevertheless, there may be occasions in your course where you simply need to commit bits of information to memory. If and when this occurs, you need some helpful strategies.

Basically there are two ways to mechanically transfer information to long-term memory. The first way is to simply repeat the information aloud over and over again until it gets "burned" in. If, for example, you have to memorize a list of items, it helps to say the first two to three items over and over until you can recall them. Then add the fourth item and recite items one, two, three, and four until you have them memorized. Then add the fifth item and so on, always going back to the first and following items, until you have the entire list memorized. This is the difficult way, but it does eventually work if you are willing to put in the considerable time and effort required.

Fortunately, there is an easier way. The easier way takes a little creative thought on your part, but it will reduce the time needed and will make the information easier to recall for longer periods of time. This easier way takes advantage of the fact that transferring information to long-term memory also occurs when you mentally link the information to be learned to some item or items of information that are already in long-term memory. When you can mentally create such a link, the new information gets transferred to long-term memory instantaneously and, in many cases, can be recalled for an indefinite period of time.

For example, have you ever unsuccessfully tried to recall the name of someone you just met? What was his name? Was it Tony, Tim, or Tom? A way to make the name easy to recall is to link the image of the person and his name to the image of someone with the same name whom you already know, while the new person's name is still present in your short-term memory. So when the new person says, "Hi, my name is Tim Jenkins," you should ask yourself, Whom do I already know named Tim? Assuming you can "pull up" the image of a person from your past with the same name, you will instantaneously form a link between the new person's face and name, and the face and name of the familiar person. So now, when you need to recall the name of the new person, you can think of the familiar person, his name, and, consequently, the name of the new person. How can this idea of linking new information to information already in your memory work in biology?

Suppose, for example, you suspect that on your next exam your instructor is going to ask you to recall, in order, the taxonomic levels of kingdom, phylum, class, order, family, genus, and species (see Chapter 6).

What images and words do you already have in memory that might help? How about trying to imagine *K*ids *P*laying *C*atch *O*n *F*ine-*G*rained *S*and? Note that this image, and these words, give us the order K, P, C, O, F, G, and S, which is precisely the order needed to recall the taxonomic levels. Words, phrases, or sentences that are associated with difficult-to-recall words, are called **mnemonic devices.**

Another mnemonic device that is commonly used in biology is CHNOPS. CHNOPS stands for *C*arbon, *H*ydrogen, *N*itrogen, *O*xygen, *P*hosphorus, and *S*ulphur, the elements that make up the vast majority of molecules found in living things.

Although it obviously takes some time to find or create mnemonic devices like these, they are extremely helpful in the recall of information. Their use makes memorization a lot easier and can save you a lot of time. Keep in mind, however, that before you decide to attempt to commit such information to memory, make sure that doing so will pay off. Once again, the key point is that you must know what kind of exam your instructor will produce before you know how to study for it.

WRITING A SCIENTIFIC PAPER

In addition to exams and homework problems, you may be required to prepare and submit one or more scientific papers during the semester. Writing forces one to reflect on past experiences and on partially understood ideas and to synthesize them into clear wholes. Thus, writing provides an excellent force for learning. Perhaps the most common scientific paper is the laboratory report. This report is similar to the scientific research paper that scientists prepare and publish in scientific journals. The laboratory report should be structured in the same way as a scientific research paper in that it should include sections devoted to the causal question raised, the alternative explanations proposed, the experimental procedure used, the predicted results, the actual results, and a discussion of the results, including a statement of the conclusion(s) drawn. Preparing such a report will be particularly challenging for students who are unfamiliar with these elements of scientific thinking. What follows are the more specific guidelines the teaching staff at Arizona State University gives to our incoming biology students to help them in preparing the laboratory report.

Guidelines for Laboratory Reports

During the semester, you will be required to submit a detailed laboratory report based on one or more of your investigations. The following information may help you understand the format of the paper. Note that not only will you be graded on content (for example, creativity, reasoning, experimentation) but also on style (grammar, spelling, neatness). The report should be divided into the following sections:

Causal Question

In this section, you should state the causal question you are addressing, that is, a question that asks why something happened: What is/are the cause(s)? You should include an introduction that consists of any background information your reader might need and a discussion of why the question is important/interesting scientifically.

Alternative Hypotheses (Explanations)

In this section, you will present the alternative hypotheses (at least two) that you will be testing. Be sure your proposed explanations can be tested using the available facilities. For example, suppose you ask, Why does water rise in plants? and hypothesize that water rises because there are little pumps in the roots, and/or there are one-way valves in the stems. These are reasonable hypotheses. To test them, you must be able to imagine and conduct an experiment that yields expected results (predictions) and actual results (your data). Your hypotheses must be ones that are testable.

Experimental Procedure

This section will describe what you did to test your proposed explanations. You should include a diagram of your setup and enough description so that someone unfamiliar with your experiment could repeat it. State your independent variable (the factor that you manipulate) and your dependent variable (what you are measuring) for each experimental design. Be very sure that you have designed controlled experiments (experiments in which there is only one independent variable) and that you have enough data to differentiate random variations from "real" variations, that is, you may need to repeat your experiment several times.

Expected Results (Predictions)

Your expected results (predictions) are derived from your proposed explanations and your experimental design. To generate a prediction, one assumes, for the purpose of investigation, that the explanation is correct. The prediction may be stated as part of an If-Then statement. The If portion is essentially a restatement of your proposed explanation. There is an explicit And portion, which represents your experiment, and a Then portion that states the results you expect to find if your proposed explanation is correct.

Example: *If* there are pumps in the roots of plants that allow water to rise (proposed explanation), *And* we cut off the roots of a group of experimental plants while not cutting off the roots of a second group of plants (experiment), *Then* we would expect that those plants with intact roots would show a greater water rise than those plants without roots (prediction).

Important Note: The If-Then statement is simply a convenient was to illustrate the relationship among a proposed explanation, experiment, and expected results. A hypothesis does not have to include the word If and a prediction does not have to start with Then. You cannot differentiate hypotheses from predictions just by looking for these cue words, because they are frequently omitted. You must understand the difference between them.

Actual Results

In this section, you will present the actual results of your experiment. Your data should be quantitative, and should be presented in tables and/or graphs. Be very careful to clearly label the axis on graphs, and columns and rows on tables. You should also include a verbal discussion of major results.

Discussion and Conclusion

In this section, you will identify trends in your data and discuss whether these trends agree or disagree with predictions derived from your proposed explanations. You may discuss any suggestive qualitative observations, explain any anomalous results, and suggest possible improvements in your experimental design. In conclusion, you should decide whether to accept or reject your hypotheses based on the results of your experiments. Do your actual results agree with your expected results? If so, then the hypotheses have been supported. If you decide to reject your proposed explanations, you may be able to suggest post hoc additional explanations at this time. Do your results suggest any further investigations of interest? If so, briefly discuss them.

TIPS ON REVISING YOUR PAPER

Once you have written your first draft, it is time to read it very carefully and try to make improvements. Your central goal is to put yourself in the reader's place and make the paper as easy to understand as possible. Very few writers are able to get it right the first time, so the revision process is very important. Here are some specific things to look for:

Delete Unnecessary Prepositions

Prepositions are words such as *by, to, in,* and *from* that are placed before nouns.

The results indicated a role of xylem tubes in water transport to the leaves of the plant.

The sentence contains four prepositional phrases painfully strung together. Consider this revision that eliminates three of them.

The result indicated that xylem tubes transport water to the plant's leaves.

Thanks to the deletion of unnecessary prepositions, this sentence is much easier to understand.

Use Strong Verbs

Verbs are words used to express an action. Scientific writing often lacks much action in the first place, so use as strong a verb as possible. Consider this sentence that contains two weak verbs:

The transport of blood is dependent on the contracting of the heart which is a muscle.

The verbs *is dependent* and *is* are pretty dull. Let's go for something slightly stronger and simpler:

The transport of blood depends on contraction of the heart muscle.

Do you see a way to clean up the sentence even more? What about some unneeded prepositions such as *of blood* and *of the heart*? If we eliminate these, we arrive at:

Blood transport depends on heart muscle contraction.

Is this sentence easier to understand? If so, the revision has been a success.

Start by Telling the Reader Who Did It

Consider the following pairs of sentences:

The dog was experimented on by Pavlov.
Pavlov experimented on the dog.

Crabs were collected from six offshore areas.
Jones collected crabs from six offshore areas.

Little is known about the influence of excessive light.
We know little about the influence of excessive light.

In the present study, the food intake of each species was recorded, and the size of the teeth were measured.

In the present study, Jones measured and recorded the food intake and teeth size of each species.

The findings were interpreted as indicative of...

I interpreted the findings as indicative of...

Notice that something is either missing from the first sentence of each pair or that something comes late in the sentence. That something is the "who," that is, who collected the crabs? Who knows little about the influence of excessive light? Scientific writers often fall into the trap of making the reader guess who did it. The less guessing, the easier it will be for the reader to understand. So, when you revise a sentence, do not forget to tell us who and tell us as soon as possible.

Make the Organism the Actor

Compare the following pairs of sentences:

Studies show that the diets of mice vary with the season (Lawson, 1992).

Mice diet varies with the season (Lawson, 1992).

Increases in pH of the soil induced greater growth rates of plants.

Plants grew faster with increases in soil pH.

Increases in salinity above the normal concentration of sea water inhibited the hatching of brine shrimp eggs.

Brine shrimp egg hatching was inhibited by salinity increases above the normal concentration of sea water.

Notice how the second sentence of each pair moves the organism to the start of the sentence and makes the organism the actor. Doing so makes it easier to understand because the reader does not have to keep one or more thoughts in mind while reading the rest of the sentence to discover what it is about.

Make Your Sentences and Paragraphs Flow Together

Read the following passage with an eye toward the flow from one sentence to the next and from the first paragraph to the second:

Harvey's blood circulation theory leads to an interesting prediction that can be used to further test his theory. If a person handles a lot of garlic, his or her breath will soon smell garlicky. Garlic molecules pass through the skin and into the blood and then circulate, eventually ending up in the mouth. A tall man with bare feet was placed horizontally through a hole in a wall, his feet in one room and his head in another. Garlic was rubbed on the bottom of his feet.

In just a few seconds, an observer standing by the man's head smelled garlic on the man's breath. The circulation theory is supported. Many other experiments and observations support the circulation theory.

Do the sentences flow easily from on to the next? Is the second paragraph linked to the first? Most likely your answer to these questions is no. Now read the following revised passage. Note the underlined linking words and phrases that have been added to smooth the transitions and provide links:

Harvey's blood circulation theory leads to an interesting prediction that can be used to further test his theory. It was known at the time that if a person handles a lot of garlic, his or her breath will soon smell garlicky. This observation suggests that garlic molecules pass through the skin and into the blood and then circulate, eventually ending up in the mouth. To test this idea, a tall man with bare feet was placed horizontally through a hole in a wall, his feet in one room and his head in another. Then garlic was rubbed on the bottoms of his feet.

The result of the experiment was that in just a few seconds, an observer standing by the man's head smelled garlic on the man's breath. Because this is the result predicted by the theory, the circulation theory is supported. Of course, many other experiments and observations since Harvey's time also support the circulation theory.

I hope you agree that the sentences of these two paragraphs, and the paragraphs themselves, flow better thanks to the addition of the connecting words and phrases.

In general, in revising for flow, you should aim to use appropriate links and transitions. Always remind the reader what has come before and help the reader anticipate what will come next. Use such expressions as *thus* and *in contrast to, moreover, because, although, however, then, therefore, even so, in addition to,* and *in spite of* whenever appropriate. Also, do not be afraid to repeat an important idea, using slightly different words or to summarize your main ideas.

Delete Teleological and Anthropomorphic Expressions

As far as we know, organisms other than humans are not conscious of any purposes for their behaviors. For example, a young mouse does not eat with the conscious intent of growing bigger. Likewise, such organisms as giraffes do not purposefully evolve long necks "in order to reach the leaves in tall trees." Attributing a sense of purpose to other living things is being **teleological** (showing a design or purpose) and should be avoided. For instance:

> Squirrels collect extra food in the fall so that they can survive the long winter.
>
> Cheetahs become fast in order to capture more prey.

More biologically appropriate revisions would be as follows:

> The extra food collected by squirrels in the fall enables them to survive the long winter.
>
> The speed of cheetahs enables them to capture more prey. (Presumably in the past the slower cheetahs died while the faster ones survived; see Chapter 13).

Your writing should also avoid giving human characteristics to nonhuman entities, called **anthropomorphism.** The following sentence is anthropomorphic and should be changed:

> Nature's great experiment in the hot and dry southwestern deserts has resulted in a diverse group of plants and animals that are pleased with their lives in the sun.

Here is a more biologically acceptable version:

> A diverse group of plants and animals are well suited to living in the hot, dry climate of the southwestern desert.

Incorrect Usage

There are a few commonly used words that are often used incorrectly in scientific writing:

between/among If you are writing about two things, use *between*: The results do not allow us to choose between the two hypotheses. If you are writing about three or more things, use *among*: The results do not allow us to chose among the three hypotheses.

which/that If the words that follow the *which/that* are crucial for sentence meaning, use *that*:

The mice that had salt in their diet grew larger. On the other hand, if the words that follow *which/that* can be deleted, or easily set off with commas, then use *which*: Protein molecules, *which* are often used as enzymes, are manufactured using directions found in the cell nucleus.

If you are like most people, you will probably discover that you have often used the word *which* when you should have used the word *that*. So, when you revise, go on a "which hunt" and help free your writing of excess *whiches*.

effect/affect Here are two words that can be a real problem. *Effect* can be used as a noun or a verb while *affect* is used only as a verb. *Effect* means "result or outcome" when used as a noun:

What is the *effect* (result or outcome) of using fertilizer for the growth of plants?

Effect as a verb means "to bring about:"

What changes in plant growth will fertilization *effect* (bring about) in plants?

The meaning of *affect* as a verb is only slightly different and means "to influence:"

How will fertilizer *affect* (influence) the growth of plants?

I know of no easy way to keep these words and meanings straight, so simply write a sentence and use either *effect* or *affect;* then match it against their definitions in the dictionary.

Remember that the word *data* is plural, for example: The *data* are very interesting. *Datum* is the singular form and is seldom used.

Have a Friend Read Your Paper

There may be a biologist out there somewhere who can create a well-written paper with no mistakes in it on the first try. But I have never met that person and do not know anyone who has. Biologists I know write and revise several times. But even then they realize that the paper is not finished and ready to be submitted for publication. Instead, they ask a few friends to read the paper and offer suggestions for improvement. Other people read from different perspectives and can point out unclear and awkward sentences or paragraphs and greatly assist the revision process. You may not have the opportunity to have one or more friends help, but if you do, take it. One further point, do not argue with your friend about his or her comments. Assess the value of the comments and accept the ones you think are helpful and ignore the ones you think are not.

TERMS TO KNOW

anthropomorphism
mnemonic device
teleological

CHAPTER 2

THE SCIENCE OF BIOLOGY

GETTING FOCUSED

- *Understand the nature of hypotheses and theories and how they are tested.*
- *Distinguish hypotheses from predictions.*
- *Distinguish pure from applied science.*

The purpose of this book is to help you become a successful student in college biology. To be successful, you will need to do more than memorize some biological facts. You will need to gain insight into the nature of biology as a science, into the way biologists think, and into the major theories that modern biologists use to understand the living world.

The key to unlocking the world of the biologist is in the thinking patterns biologists use to explore nature. If you have already acquired those thinking patterns, your job as a biology student will be fairly easy. If not, your job will be more difficult but by no means impossible, because this book will help you acquire them.

WHAT IS SCIENCE?

Science is our attempt to understand the objects and events we experience in nature. People develop understanding about things they experience by asking questions and finding answers. What is life? What happens to frogs in the winter when their pond is frozen over? Why are there so many different kinds of plants and animals? What causes diseases such as AIDS? In attempting to find answers to questions such as these, one is doing science. Because all of these questions involve living things, finding answers to them involves you in doing biology—the science of life.

THE NATURE OF SCIENTIFIC THINKING: THE EARLY RISER

But how do you do biology? How do you answer questions about life? How do you answer any question? Consider the following:

A few years ago, the behavior of my one-year-old son raised a question. He was waking up at around 5:00 A.M. As far as his parents were concerned, this was way too early. Why was he waking up so early? The problem was to discover the cause so that something could be done to get him to sleep longer.

Because the boy's awakening occurred in the summer, when the sun was streaming through the window early, his parents thought that perhaps the light was awakening him—a hypothesis. A **hypothesis** is a tentative (not yet tested) explanation for some experience. A hypothesis is a possible answer to the causal question raised. In this case, it is a tentative answer to the causal question: What caused the child to wake up so early? A second hypothesis was that the child was hungry and his hunger had awakened him. Although other hypotheses could be suggested, these seemed the most likely to his parents. Generating hypotheses is an important first step to answering a causal question. The next step involves testing the hypotheses to find out which one is best. How is this done?

To test any hypothesis, one must first determine what would happen if the hypothesis is right. In other words, if the hypothesis is correct, then what will you predict will happen under future conditions? Testing a hypothesis requires thinking that takes the if-then form, which is referred to as a process of **deduction.** Making these deductions is a crucial step when we cannot directly observe a cause, thus can only observe its predicted consequences. The if-then deductive thinking used to determine correctness of the present hypotheses looks like this:

Hypothesis: *If*...the sunlight coming through the window was awakening the child...

and...the sunlight is blocked with a heavy cover over the window...

Prediction: *then*...he will awaken later.

On the other hand,

Hypothesis: *If*...his hunger was awakening him...

and...he is fed an additional bottle of milk at midnight...

Prediction: *then*...he will awaken later.

The result of making a deduction is a prediction. A **prediction** is simply a statement of how you expect the experiment to turn out if the hypothesis is correct. This is not the same sort of prediction as when we say, for example, "I predict it will rain tomorrow." Such predictions arise not by deduction but primarily by guessing.

What remains to be done to test a hypothesis is simply to compare the predicted result with what in fact happens (i.e., the evidence) when you do the experiment. If what is predicted actually happens, then you have supported your hypothesis. For example, if you place a heavy cover over the window and the child awakens later, then you have supported the sunlight hypothesis. If what happens is different from what was predicted, then the hypothesis has not been supported. For example, if the child is fed just before going to bed but he still awakens at 5:00 A.M., then you have not supported the hunger hypothesis. When what is predicted does not happen, you must conclude that either something is wrong with your hypothesis or something is wrong with the way you did your experiment. This last phase of trying to answer a question is sometimes called the test phase because the purpose is to test (either support or not support) the hypotheses that have been advanced. Figure 2.1 depicts the hypothetico-deductive thinking pattern used to test the sunlight hypothesis.

Hypotheses Are Neither Proven nor Disproven

Notice that I just said that the purpose of the test phase in hypothesis testing is to either *support* or *not support* the hypotheses that have been advanced. You might want to read the word *support* as equivalent to the word *prove* and the words *not support* as equivalent to *disprove.* But this would be stating your conclusions too strongly. In fact, hypotheses can neither be proven nor disproven with any finality. Some doubt about the truth or falsity of hypotheses must always remain.

The reason that supportive results do not prove a hypothesis correct is because hypotheses, being the products of human imagination, are potentially unlimited in number, and any two or more may give rise to the same prediction(s) (that is, when hypothesis A predicts *x,* and hypothesis B predicts *x,* the observation of *x* cannot tell you whether it was A or B that led to *x*). For example, suppose you generate the hypothesis that the sun orbits the Earth. Such a hypothesis leads to the predictions that the sun will rise on one horizon, cross the sky, and set on the opposite horizon. Consequently, when such a set of events are observed, the "sun-orbit" hypothesis gains support. But it has not been proven, because some other hypothesis could give rise to the same set of predictions (for example, the Earth rotates on an axis).

But why are hypotheses not disproven? Consider the following example. Suppose on a walk in a park you observe the two nearly identical trees shown in Figure 2.2 on page 38. Tree A has tall grass growing under it while tree B has nearly none. Let us generate several hypotheses to account for the difference in the grass growth beneath the trees (e.g., tree B provides too much shade, tree B drops grass-killing fruit, children trample the grass under tree B, the grass under tree A is shade tolerant). Now let us test one of these hypotheses. The first hypothesis leads to the prediction that the grass should grow if the branches are cut off to permit more sunlight to reach the ground near the base of tree B. Suppose we conduct the experiment and after several weeks we discover that the grass does not grow. Have we, therefore, disproven the hypothesis? Recall that the logic of the situation reads as follows: The

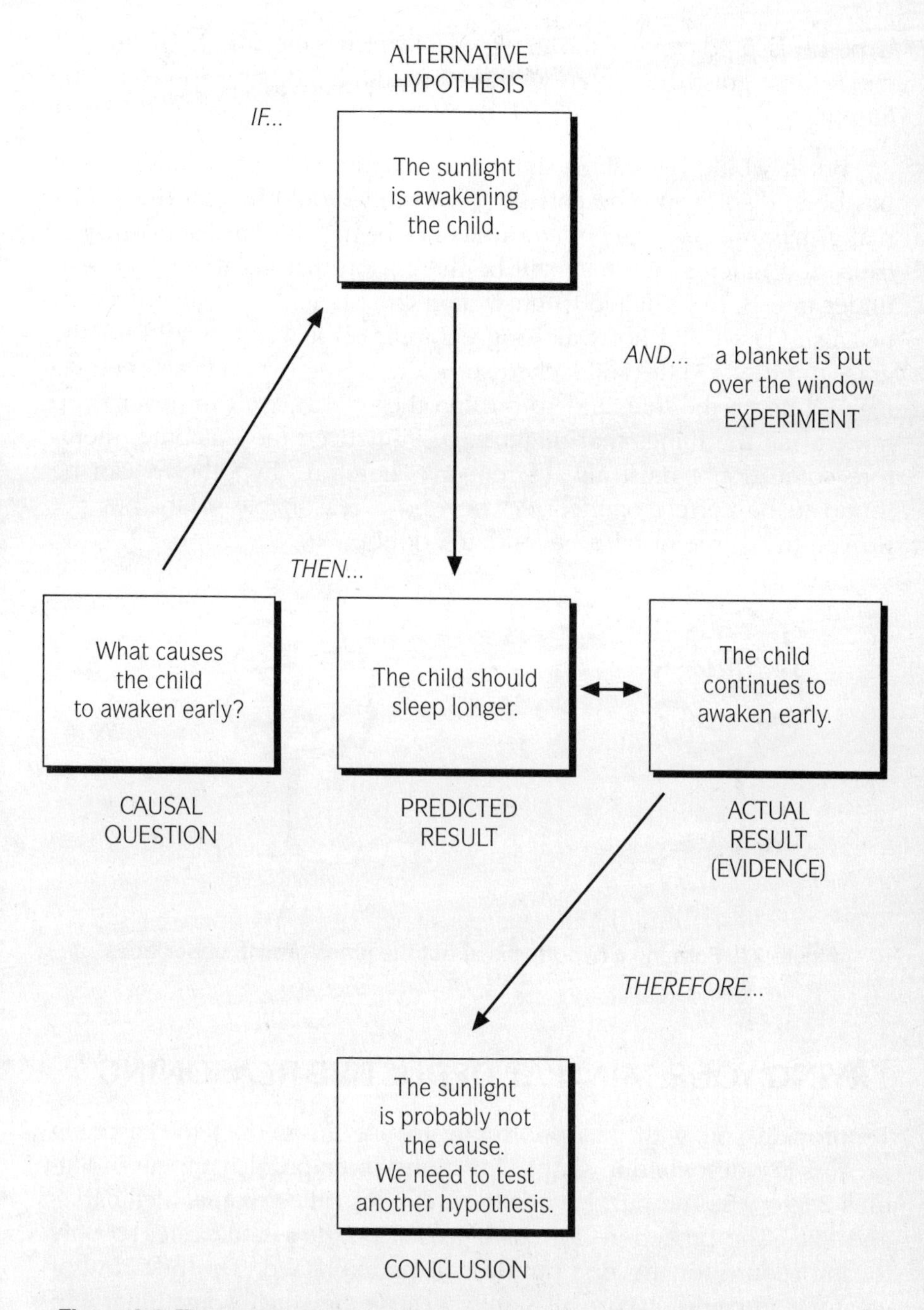

Figure 2.1 The pattern of hypothetico-deductive thinking used to test the alternative hypothesis that the sunlight is causing the child to awaken early.
(From *Biology: A Critical Thinking Approach,* Student Readings, by Anton E. Lawson © 1994 by the Addison-Wesley Publishing Company.)

hypothesis predicts that the grass should grow after the branches are cut, but the grass did not grow; therefore, the hypothesis appears to be false.

But it would be a short-sighted investigator who concludes that it has been disproven. The correct conclusion should be that the explanation has *not been supported*. It has not been falsified/disproven, because too much shade may still be the reason that grass did not grow under tree B, but it failed to grow after the branches had been cut off because (1) we did not wait long enough, (2) it was too cold for the grass to grow, (3) the soil lacked sufficient water, (4) no grass seed remained under the tree, and so on. In other words, we can never keep track of all the things that might have influenced the outcome, therefore some doubt must always remain. Therefore, hypotheses can be found to be correct or incorrect beyond a reasonable doubt but not proven to be true or false beyond any doubt.

Figure 2.2 Forming a hypothesis about the grass growth under trees.

TRYING YOUR MIND AT DEDUCTIVE REASONING

To provide you with an opportunity to use the deductive reasoning process just described in simple situations, three puzzles appear in Figures 2.3 to 2.5. The puzzles involve creatures called Skints, Mellinarks, and Pilts. The procedure for solving each puzzle is the same, so only the procedure for the first puzzle will be explained. The first row of creatures in Figure 2.3 are all Skints because they have something(s) in common. None of the figures in the second row are Skints because they lack that/those something(s). Based on this information, your task is to figure out which of the figures in the third row are Skints. Once you think you have solved the Skints puzzle, go on to the others.

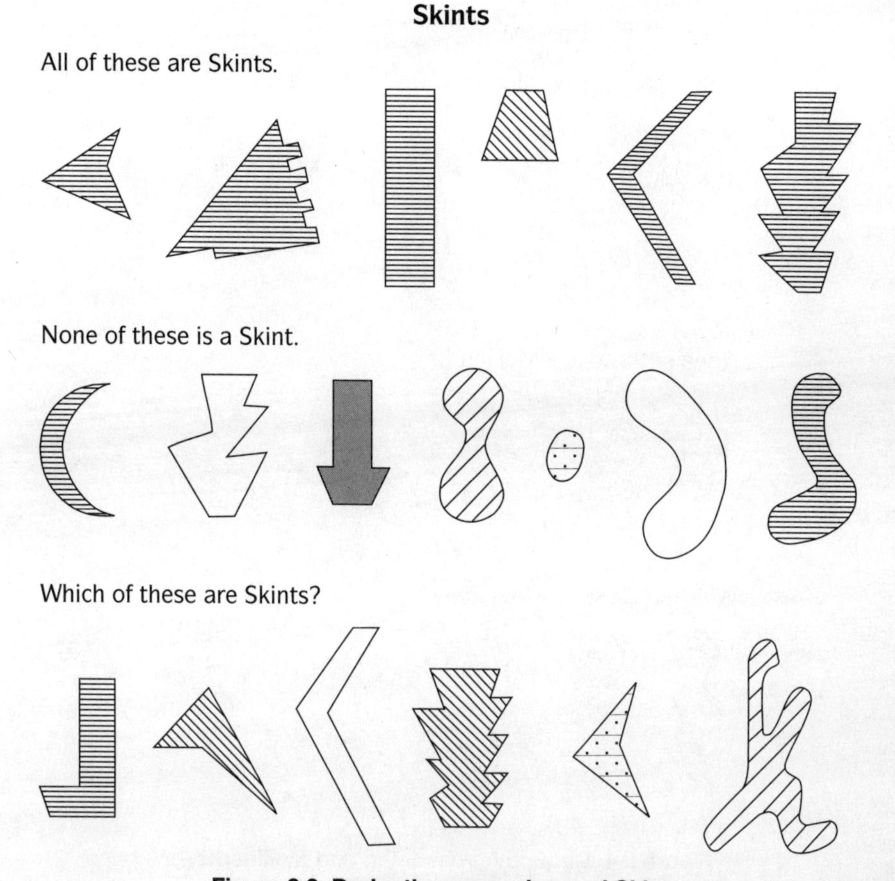

Figure 2.3 Deductive reasoning and Skints.

When you are finished, compare your answers with a friend or two and, more importantly, the reasoning patterns you used to arrive at these answers. Make sure to identify the ideas and the deductions that you developed. Name some of the ideas that you rejected and write a sentence or two to summarize the reasoning that led to their rejection. For example, the following argument states why the idea that "Mellinarks are creatures defined solely by the presence of a tail" is insufficient and must be modified or rejected:

Idea: *If*...Mellinarks are creatures defined solely by the presence of a tail...

and...I examine the non-Mellinarks in row 2...

then...none of the non-Mellinarks in row 2 should have a tail.

Mellinarks

All of these are Mellinarks.

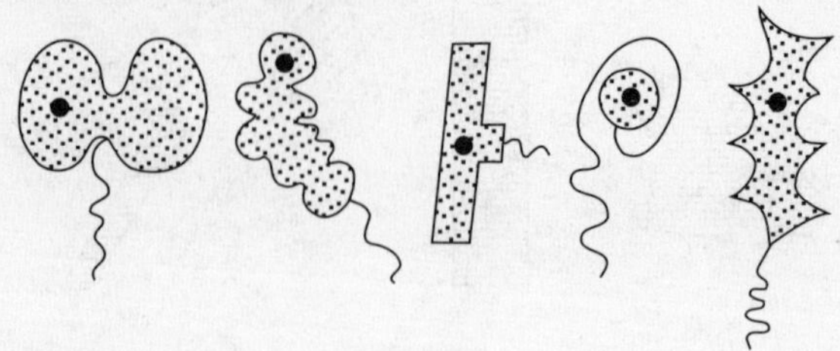

None of these is a Mellinark.

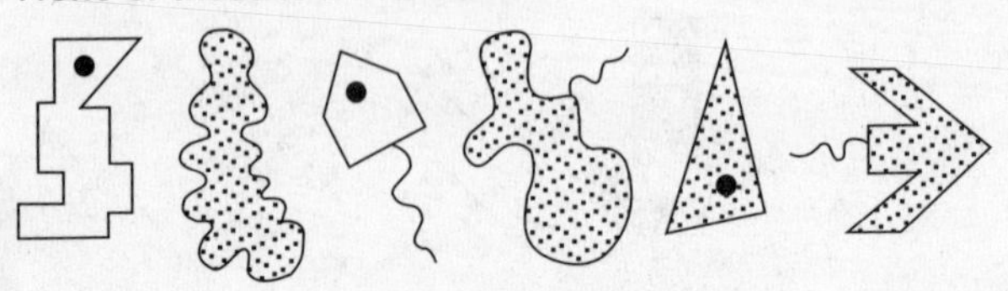

Which of these is a Mellinark?

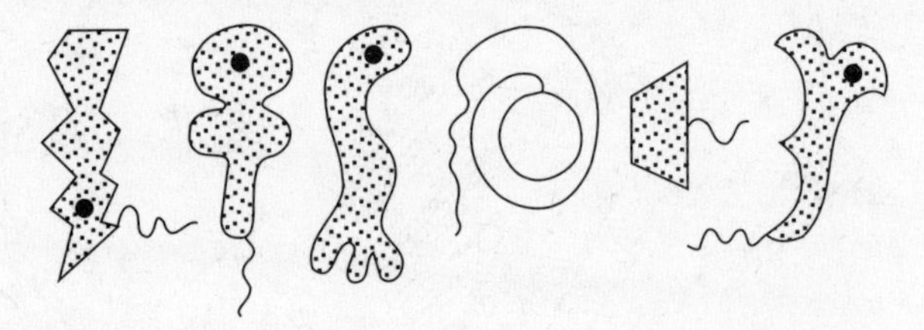

Figure 2.4 Deductive reasoning and Mellinarks.

Result: *But*...the first, third, fourth, and sixth creature in row 2 have tails.

Conclusion: *Therefore*...the idea that Mellinarks are creatures defined solely by the presence of a tail must be rejected. I need to generate another idea, and so forth.

Reasoning and Concept Acquisition

If you were successful in using the deductive reasoning pattern to discover what Skints, Mellinarks, and Pilts are, we can say that you have acquired three new concepts. A **concept** is defined as a mental creation that consists of some identified pattern linked to a term or terms.

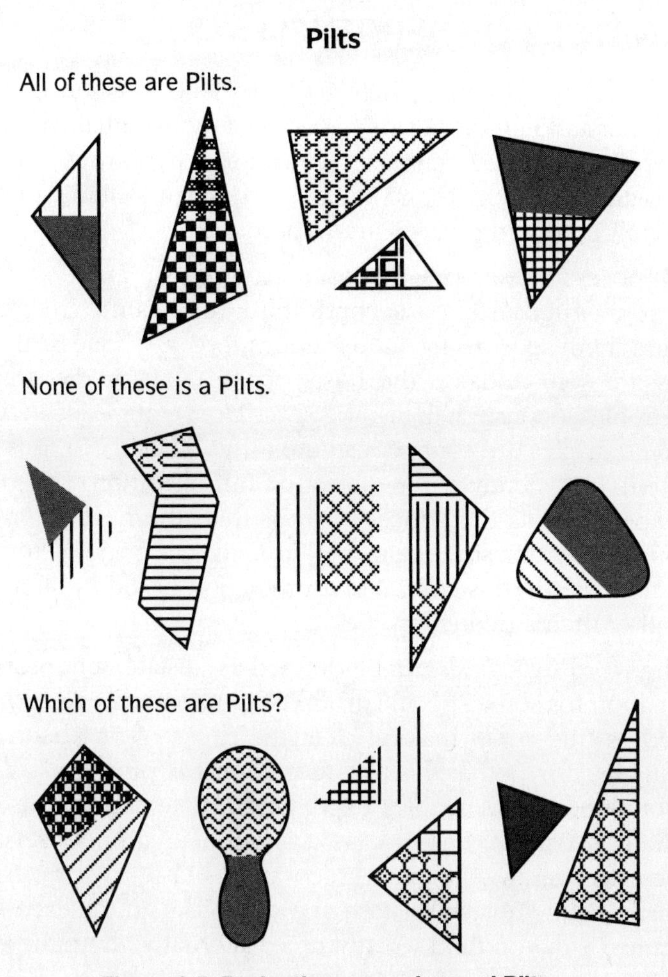

Figure 2.5 Deductive reasoning and Pilts.

For example, the pattern for the term *Mellinark* is a drawing on paper that consists of one big dot, lots of little dots, and a tail.

If you were able to select creatures 1, 2, and 6 of row 3 in Figure 2.4 as Mellinarks, we can say that you have acquired the concept of Mellinarks. The important point as far as learning biology is concerned is that to be successful you will need to acquire a lot of new concepts (for example, homeostasis, osmosis, natural selection). As you can see, acquiring concepts requires deductive reasoning and several related reasoning patterns. Therefore, much of the next three chapters will be devoted to helping you become a better reasoner.

THE NATURE OF HYPOTHESES

Because hypotheses play such a crucial role in science, obtaining a clear understanding of that role is absolutely essential. Hypotheses are not merely educated guesses. Generating hypotheses does require background knowledge and it does require an element of guessing, but not all educated guesses are hypotheses.

Suppose, for example, that you taste a green apple and discover it tastes sour. After tasting a second, third, and fourth green apple, you also find them sour. So from this "education" you "guess" that all green apples are sour and, on the basis of this, you predict that the next green apple you taste will also be sour. Does your educated guess that "all green apples are sour" constitute a hypothesis? No, it is not an explanation, it is merely a generalization (an induction) based upon limited experience. Is the educated guess that "the next green apple will be sour" a hypothesis? Again, no. Instead it is better referred to as a prediction. But in this case, it is a prediction based upon a generalization rather than a deduction.

The word *hypothesis* can be defined as "a statement proposed as an explanation for some specific group of phenomena." An *explanation* is defined as "the act of making clear the cause of or reason for" something. Thus, a hypothesis is a statement that is proposed as a possible, but not directly observable, explanation, a tentative cause for some specific observation. In this case it is a tentative nonobservable cause for the observation that some green apples taste sour. Perhaps the apples lack sugar molecules. Perhaps they contain an excess of "sour" molecules. Can you think of any other alternative hypotheses?

Generating hypotheses consists of studying the facts and devising possible explanations for them. Obviously, doing so requires some education and some guessing about causes, the guesses coming not from induction or deduction, but from a creative process called *abduction*. **Abduction** involves sensing ways in which the current situation is somehow similar (analogous) to other known situations and in using this similarity as a source of hypotheses. Perhaps you know that sugar molecules make candy and cookies sweet, so it seems reasonable to borrow this idea and use it to explain the lack of sweetness in the green apples. Thus, the statement that green apples are sour because they lack sugar molecules is a hypothesis. Of course, it may or may not be "true."

Finally, the good scientist does not merely consider one possible cause, but considers as many alternative causes as he or she can think of and then sets out to devise ways of testing the alternatives by deducing their consequences and comparing these with evidence, as depicted in Figure 2.1.

Now think back to how you solved the Skints, Mellinarks, and Pilts puzzles. Suppose on the Mellinark puzzle you initially came up with the idea that the Mellinarks are creatures defined solely by the presence of a tail, or, on the Pilts puzzle, that Pilts are triangles. Where did these ideas come from? The answer of course is that observable features such as tails and triangles are present in the creatures shown in the first row. This means that the ideas came from observation and induction rather than from imagination and abduction. The reasoning used to generate and test your ideas about the Skints, Mellinarks, and Pilts was, therefore, inductive-deductive; whereas the reasoning used to generate and test scientific hypotheses is abductive-deductive (sometimes referred to as hypothetico-deductive). You might be quite comfortable with the inductive-deductive reasoning, but hypothetico-deductive reasoning may be rather new to you and may seem a bit strange. Indeed it is a bit strange. In essence, it requires that you generate imaginary explanations and assume for the time being that they are correct, just so that you may find them incorrect.

Pure and Applied Biology

What motivates someone to try to discover the causes of such events as grass growing under one tree and not under another or a child waking up at 5:00 A.M.? When the motivation is simply one's curiosity, the process is referred to as **pure science.** One the other hand, when the researcher is motivated by practical goals, such as trying to get grass to grow under trees or getting a child to sleep later, the process is referred to as **applied science.** Importantly, whatever the motivation, the thinking patterns involved in both pure and applied science are the same. Another important point is that even though many of the past and present questions raised by biologists have been and are motivated simply by curiosity, the knowledge gained has led and will continue to lead to applications down the road. These applications include methods of increased agricultural productivity, disease prevention and cure, resource management, waste management, and population control.

THE NATURE OF THEORIES

Biology consists not only of its methods of inquiry but of its theories as well. The term **theory** refers to combinations of statements that function together to explain a phenomenon or set of related phenomena. A theory may or may not represent an adequate explanation. Many theories of the past seemed adequate at the time but have subsequently been rejected or modified. Nevertheless, they remain theories. To determine whether or not a theory is a "good" one, its basic statements must be tested as discussed above. It is by no means an understatement to say that the central purpose of modern biology is to generate and test comprehensive theories about life.

Perhaps the best way to provide you with a sense of how the term *theory* is used is to offer a few examples. Below are the major statements (sometimes referred to as postulates, basic premises, or fundamental assumptions) of two major theories in biology.

POSTULATES OF GREGOR MENDEL'S THEORY OF INHERITANCE

To explain how characteristics are passed from parent to offspring

1. Inherited characteristics are determined by particles called factors.
2. Factors are passed from parent to offspring in the gametes (egg and sperm cells).
3. Individuals have at least one pair of factors for each characteristic in each body cell except the gametes.
4. During gamete formation, paired factors separate. A gamete receives one factor of each pair.
5. There is an equal chance that a gamete will receive either one of the factors of a pair.
6. When considering two or more pairs of factors, the factors of each pair pass independently to the gametes.
7. Factors of a pair that were separated during formation of the gametes unite randomly during fertilization.
8. Sometimes one factor of a pair dominates the other factor, so that it alone controls the characteristic (dominant/recessive).

POSTULATES OF CHARLES DARWIN'S THEORY OF NATURAL SELECTION

To explain how organisms adapt to their environments

1. Populations of organisms have the potential to increase rapidly.
2. In the short run, the number of individuals in a population remains fairly constant because the conditions of life are limited.
3. Individuals in a population are not all the same; they have variations (variable characteristics).
4. There is a struggle for survival, so that individuals having favorable characteristics survive and produce more offspring than those with unfavorable characteristics.
5. Some of the characteristics that enable survival and reproduction are passed from parent to offspring (i.e., they are heritable). Therefore, there is a natural selection for certain favorable characteristics.
6. The environments of many organisms have been changing for vast periods of time.
7. Natural selection causes the accumulation of favorable characteristics and the loss of unfavorable characteristics to such an extent that new kinds of organisms may arise.

Mendel's theory of inheritance was first proposed in 1866, and Darwin's theory of natural selection was first published in 1858. Although research on inheritance and evolution continue to this day, most of the postulates of both of these theories have withstood the test of time as the available evidence supports their validity. Consequently, the theories play very important roles in modern biological thought. Although the postulates of these theories and other theories may take on the status of "fact," the possibility of coming up with a better theory or with evidence that contradicts one or more of the postulates always remains. Therefore, absolute certainty is not attainable.

Theories play a central role in science, but you should not conclude that their role is limited to the sciences. Theories also play a central role in virtually all walks of life, including how countries govern themselves. Consider, for example, the postulates of a political theory of government put forth in 1776 in the Declaration of Independence:

POSTULATES OF A POLITICAL THEORY OF GOVERNMENT

1. All men are created equal.
2. All men are endowed by their creator with certain unalienable rights, among them life, liberty, and the pursuit of happiness.
3. To secure these rights, governments are instituted.
4. Governments derive their just powers from the consent of the governed.
5. Whenever any form of government becomes destructive of these ends, it is the right and duty of the people to alter or abolish it, and to institute a new government.
6. The new government should be based on the principles stated and its powers organized in such form as to most likely effect the safety, happiness, and future security of the people.

Although this book restricts itself to the discussion of biological theories, you should keep in mind that theories exist in all walks of life. Your understanding and success will be enhanced if you are able to see past the detail and identify and test the key postulates of these theories, whatever they are.

POSTULATES OF THE THEORY OF KNOWLEDGE ACQUISITION

1. Causal questions about nature are tentatively answered by putting forth alternative explanations that use a process called *abduction.*
2. Alternative explanations may consist of single statements that attempt to explain a single phenomenon or a group of closely related phenomena, in which case the explanations are referred to as *hypotheses.*
3. Alternative explanations may consist of several statements, called *postulates,* that taken together attempt to explain a broad class of phenomena, in which case the explanations are referred to as *theories.*
4. Alternative explanations are tested by experiments that allow the deduction of specific observational predictions from the proposed explanatory statement or statements.
5. Predictions are then compared with observational results (evidence). If the results are as predicted, the statement that led to the prediction is supported. If not, it is not supported.
6. Hypotheses and theories may be satisfactory or unsatisfactory explanations, depending upon the evidence that has been gathered in their favor or disfavor and the extent to which they fit with other established hypotheses or theories.
7. When the postulates of a particular theory are continually supported by evidence, the theory itself may become widely accepted.
8. Because the possibility of a better theory and/or contradictory evidence always remains, no theory can be considered correct in any absolute sense.

Please do not be overly concerned at this point if you do not fully understand many of these points. They are being presented at the outset merely to provide you with a general sense of the nature of biological thought processes. Although much of this may be somewhat confusing now, hopefully by the end of the book it will make perfect sense.

QUESTIONS

1. During the next day or so, be on the lookout for objects, events, or situations that raise questions about causes. For example, on a walk you might observe a spot of yellow grass in the middle of someone's green lawn and ask: What caused the yellow spot? Or you might be watching television only to have the picture flicker off and you ask: Why did the picture go off? Make a list of five such causal questions.

____

____

____

____

____

2. For one of the five questions listed above, propose two alternative answers (hypotheses).

____

____

____

____

3. Use the if… and …then reasoning pattern to generate a prediction to the test one or both of these alternatives.

____

____

____

____

4. For many students, the process of developing and testing hypotheses seems a bit odd. Instead of simply observing the world to see if your explanation is correct, you must start by assuming that the explanation is correct so that it may be shown to be incorrect! If the explanation is not shown to be incorrect, then it can be retained for the time being at least. Although this approach to learning about the world may seem backward, it nevertheless is the basic pattern of scientific thinking. Try your mind

at these two puzzles to see if you can identify this "backward" thinking pattern.

The Four Index Cards

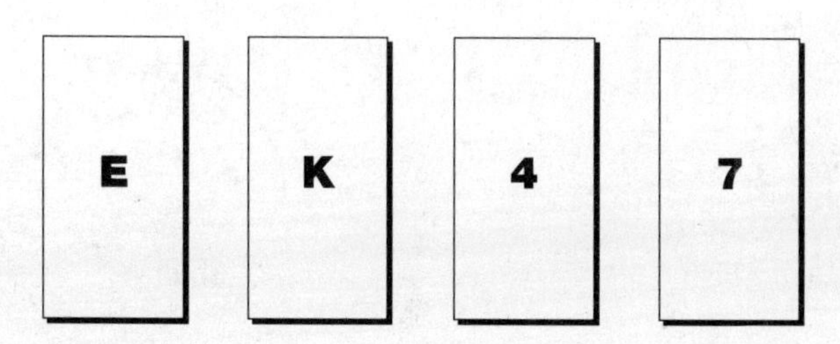

Each of the cards has a letter on one side and a number on the other side. Read the following rule:

"If a card has a vowel on one side, then it has an even number on the other side."

a. Suppose you want to test to see if this idea is correct or incorrect for these four cards. Which of the four cards must be turned over to allow the rule to be tested? Explain.

____

____

____

____

The Island Puzzle

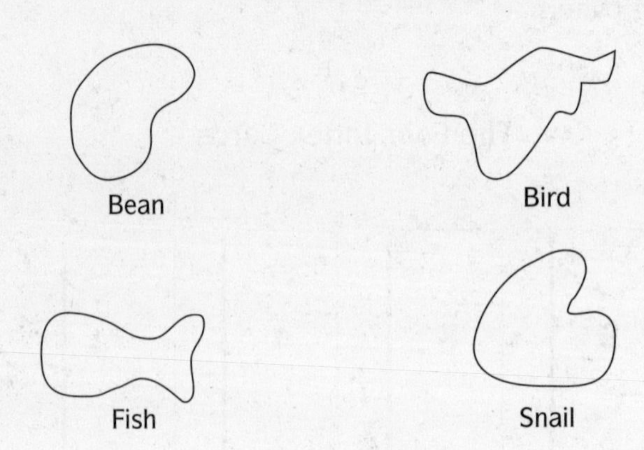

The puzzle is about four islands called Bean Island, Bird Island, Fish Island, and Snail Island. People have been traveling among these islands by boat for many years, but recently an airline started up. Carefully read the clues below about possible plane trips. The trips may be direct or they may include stops on one of the islands. When I say a trip is possible, I mean it can be made in both directions between the islands.

First Clue: People can go by plane between Bean and Fish islands.

Second Clue: People cannot go by plane between Bird and Snail islands.

Use these two clues to answer question b.

b. Can people go by plane between Bean and Bird islands?

Yes No Can't tell from the clues

Explain your answer.

____

____

____

____

Third Clue: People can go by plane between Bean and Bird islands.

Use all three clues to answer questions c and d. Don't change your answer to question b.

c. Can people go by plane between Fish and Bird islands?

Yes No Can't tell from the clues

Explain your answer.

____

____

____

____

d. Can people go by plane between Fish and Snail islands?

Yes No Can't tell from the clues

Explain your answer.

____

____

____

____

TERMS TO KNOW

abduction
applied science
assumption
causal question
concept
conclusion
deduction
evidence
experiment
fact
hypothesis
hypothetico-deductive thinking
induction
postulate
prediction
pure science
theory

CHAPTER 3

DESCRIPTIVE AND HYPOTHETICAL THINKING PATTERNS

GETTING FOCUSED

- *Identify specific scientific thinking patterns.*
- *Distinguish descriptive from hypothetical thinking patterns.*
- *Learn how thinking patterns are used to test hypotheses.*

This chapter looks more closely at the process of doing science to identify specific thinking patterns involved in the process of hypothesis testing. To succeed in biology, you will need to understand and be able to use these thinking patterns.

HOW GOOD ARE YOU AT SCIENTIFIC THINKING?

Below you will find three puzzles that have been given to students to solve. Each puzzle is followed by several typical student responses. First try to solve each puzzle, then compare your ideas with those of the students. Later you will learn what the responses reveal about how students think.

The Mealworm Puzzle

An experimenter wanted to test the response of mealworms to light and moisture. To do this, he set up four boxes as shown in Figure 3.1 below. He used neon lamps for light sources and constantly watered pieces of paper in the boxes for moisture. In the center of each box, he placed 20 mealworms. One day later he returned to count the number of mealworms that had crawled to the different ends of the boxes.

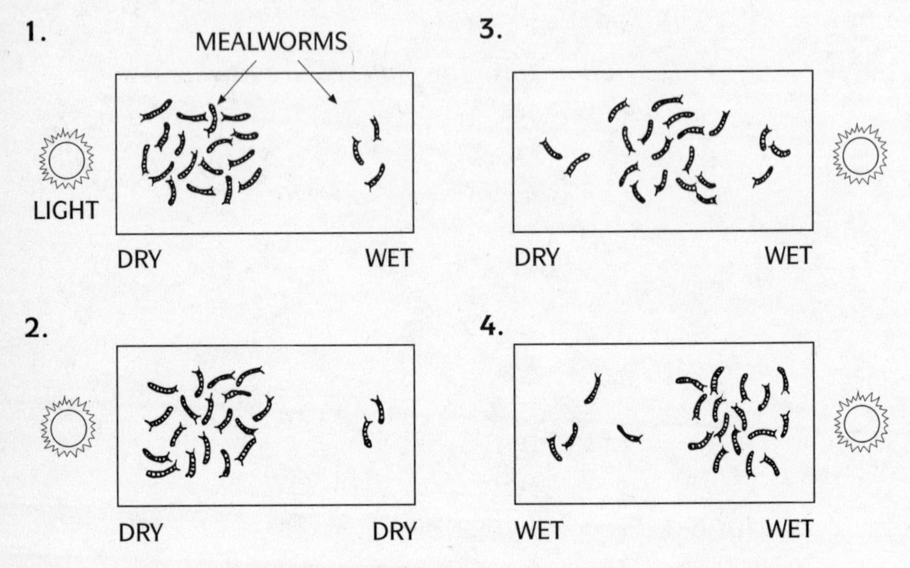

Figure 3.1 Testing mealworm response to light and moisture.

The diagrams in Figure 3.1 show that mealworms respond to (that is, move toward or away from)

a. light but not moisture.

b. moisture but not light.

c. both light and moisture.

d. neither light nor moisture.

Which of the choices seems most reasonable? Please explain your choice.

____

____

____

____

The following are typical student responses to the Mealworm Puzzle. Read these responses and compare them with your own. Look for similarities among Type A responses and among Type B responses. Look for differences between Type A and Type B responses.

Student A-1 (college freshman) This student selected d, "Because, even though the light was moved in different places, the mealworms didn't do the same things."

Student A-2 (college freshman) This student selected a because "They usually went to the end of the box with the light."

Student A-3 (high school sophomore) This student selected a "Because there are 17 worms by the light and there are only 3 by moisture."

Student B-1 (college sophomore) This student selected c "Boxes 1 and 2 show they prefer dry and light to wet and dark, Box 4 eliminates dryness as a factor, so they do respond to light only. Box 3 shows that wetness cancels the effect of the light, so it seems they prefer dry. It would be clearer if one of the boxes was wet-dry with no light."

Student B-2 (high school freshman) This student selected c because "In experiment 3, the mealworms split 1/2 wet, 1/2 dry. So it's safe to assume that light was not the only factor involved."

Student B-3 (college freshman) This student selected c because "The mealworms in all cases respond to light. However, in box 3 the worms are in the middle. This shows that the worms are attracted to the light but not as much as the situations where the dry area was next to the light. When there is no choice between wet and dry, such as in Case 4, the worms turn to the light."

Note: We might also test a box with wet at one end and dry at the other with no light to further verify the effect of moisture.

Questions

1. What similarities did you find among the Type A responses ?

___

___

___

___

2. What similarities did you find among the Type B responses?

___

___

___

___

The Algae Puzzle

A population of crabs that eats algae lives on a seashore. On the seashore, there are four kinds of algae: yellow, red, brown, and green.

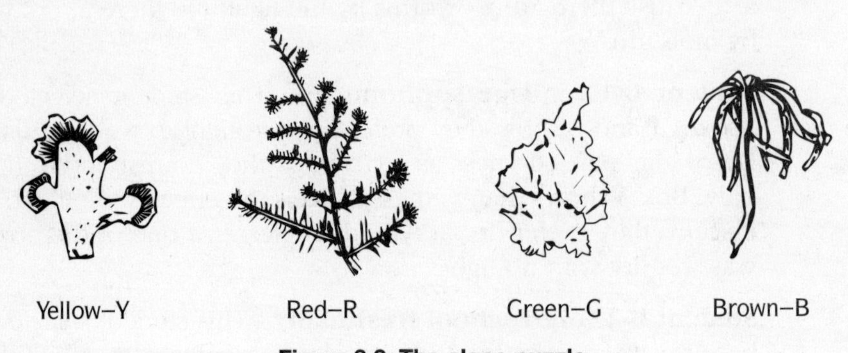

Figure 3.2 The algae puzzle.

Dr. Saltspray, a biologist, is interested in determining which of the types of algae have been eaten by the crabs. She plans to find out by cutting open the stomachs of some of the crabs and looking to see which types of algae are inside. But before she cuts open the crabs, she lists all the different algae diets she thinks it is possible to find in the stomachs. Write down each possible algae diet she might find. Use the letters Y, R, G, and B to save space.

____

____

____

____

The following are typical student responses to the Algae Puzzle. As you look them over, notice that they have been arranged roughly from worst to best. Try to discover what criteria were used for this ranking. Try to discover what distinguishes Type A from Type B responses.

Student A-1 (high school sophomore) This student listed YR, YG, YB, RG, GB, and BR.

Student A-2 (college junior) This student listed R and B, Y and R, G and Y, R and G, Y and B, and G and B.

Student A-3 (college graduate) This student listed Y, R, G, B, YR, YG, YB, RG, RB, and GB, ten combinations in all.

Student B-1 (high school junior) This student listed YRGB, YG, YB, RGB, YGB, GR, GB, and BR.

Student B-2 (college freshman) This student listed, in tabular manner:

Y, R, G, B	R, G, B	G, B	B
Y, R, G	R, G	G	
Y, R	R		
Y			

Student B-3 (college sophomore) This student listed, in tabular manner:

YRGB	RGB	GB	B	YGB
YRG	RG	G		YB
YRB	RB			YG
YR	R			
Y				

Questions

1. What criteria seem to form the basis of the ranking of responses?

____

____

____

____

2. What differences did you find between Type A and Type B responses?

____

____

____

____

The Frog Puzzle

Professor Thistlebush, an ecologist, conducted an experiment to determine the number of frogs that live in a pond near the field station. Since she could not catch all of the frogs, she caught as many as she could, put a white band around their left hind legs, and put them back in the pond. A week later she returned to the pond and again caught as many frogs as she could. Here are the professor's data:

First trip to the pond: 55 frogs caught and banded.

Second trip to the pond: 72 frogs caught; of those 72 frogs 12 were banded.

The professor assumed that the banded frogs had mixed thoroughly with the unbanded frogs, and from her data she was able to approximate (estimate) the number of frogs that live in the pond. If you can compute this number, please do so. Write it in the space below. Explain in words how you calculated your results.

____

____

____

____

Here are several student solutions to the Frog Puzzle. Please read the solutions and responses and compare them with your own. Look for key differences between Type A responses and Type B responses.

Student A-1 (college senior) Answer: 72. Explanation: "It would be 72 because that is all she was able to catch."

Student A-2 (college freshman) Answer: 115. Explanation: "55 were caught and banded, 55 with bands. She didn't catch every frog, so 55 and 60 would be 115.)"

Student A-3 (college freshman) Answer: 115. Explanation: "There were 55 frogs banded. On the second trip 72 are caught; of those 12 are banded. So 72 plus 60 new ones makes 115."

Student A-4 (college junior) Answer: about 200. Explanation:"60 frogs were caught. I'd have to try a third time (take a second sample). How big is the pond? Add 60 and 55 together? It's poor experimental data. I would come up in a third try. I'd guess 200."

Student B-2 (college sophomore) Answer: 275. Explanation: "72 – 12 = 60 were not banded. $55/x = 12/60$, $x = 275$, so the number of frogs is 275."

Student B-3 (college freshman) Answer: 330. Explanation: "You have to assume that in the week between the first and second sampling none of the banded frogs died or were born. Also, the assumption must be made that the frogs mingled thoroughly. This may not be the case, but anyway if you make all these assumptions the problem is simple. $12/72 = 1/6$ so 1/6th of the frogs have bands $55 \times 6 = 330$.

Question

1. What seem to be some significant differences between Type A responses and Type B responses?

____

____

____

____

DESCRIPTIVE AND HYPOTHETICAL THINKING PATTERNS

In reading the student responses to the three puzzles, you undoubtedly recognized that Type B answers were more complete, more consistent, and more systematic; in short, they were better than Type A answers. Type A and Type B answers are representative of two different levels of thinking.

Psychologists often characterize thinking patterns in terms of four major levels or stages. The first two are usually completed before a child is 7 or 8 years old. Therefore, only the last two are of particular interest to us; I will call them *descriptive thought* and *hypothetical thought*. Descriptive thinking patterns enable us to order and describe observable objects, events, and situations accurately. Hypothetical thinking patterns allow us to go beyond descriptions and create and test imaginary (nonobservable) explanations. What follows are some key descriptive and hypothetical level thinking patterns:

Descriptive Thinking Patterns

D1 or Class Inclusion The individual understands simple classifications and generalizations (e.g., all dogs are animals, only some animals are dogs).

D2 or Conservation The individual applies conservation thinking to observable objects and properties (e.g., if nothing is added or taken away, the amount, number, length, weight, etc., remains the same even though the appearance differs).

D3 or Serial Ordering The individual arranges a set of objects or data in serial order and establishes a one-to-one correspondence (e.g., the youngest plants have the smallest leaves).

These thinking patterns enable people to

a. Understand concepts and statements that make a direct reference to familiar actions and observable objects and can be explained in terms of simple associations (e.g., the plants in this container are taller because they got more fertilizer).

b. Follow step-by-step instructions, as in a recipe, provided each step is completely specified (e.g., can identify organisms with the use of a taxonomic key or determine the oxygen content of a water sample using a standard procedure).

c. Relate his or her viewpoint to that of another in a simple situation (e.g., a girl is aware that she is her sister's sister).

However, people whose thinking has not developed beyond the descriptive level demonstrate certain limitations. These limitations are evidenced as they

d. Search for and identify some factors influencing an event, but do so unsystematically (e.g., investigates the effects of amount of light on mealworms but does not necessarily make sure that other factors, such as amount of moisture and temperature, do not change).

e. Make observations and draw inferences from them, but do not initiate reasoning with the possible.

f. Respond to difficult problems by applying a related but not necessarily correct rule.

g. Process information but are not aware of their thinking patterns (e.g., do not check their conclusions against the given data or other experience).

The above characteristics typify descriptive level thought, that is, Type A responses.

Hypothetical Thinking Patterns

Hypothetical thinking patterns are those used in the testing of alternative hypotheses.

H1 or Combinatorial Thinking The individual systematically considers all possible relations of experimental or hypothetical conditions, even though some may not be realized in nature (e.g., generates all 15 possible diets in the Algae Puzzle).

H2 or Identification and the Control of Variables In testing hypotheses, the individual recognizes the need to take into consideration all the known factors and to design a test that keeps all but the one being tested the same (e.g., in the Mealworm Puzzle, recognizes the inadequacy of the setup using box 1, in which both the amount of light and the amount of moisture change).

H3 or Proportional Thinking The individual recognizes and interprets numerical relationships between relationships (e.g., how fast a molecule moves through a cell membrane is inversely proportional to the square root of its molecular weight; for every 1 banded frog there are 6 total frogs; therefore, for every 55 banded frogs there must be $6 \times 55 = 330$ total frogs).

H4 or Probabilistic Thinking The individual understands that nature is complex and that any conclusions or explanations must involve probabilistic considerations (e.g., in the Mealworm Puzzle the ability to disregard the few mealworms in the "wrong" ends of boxes 1, 2, and 4; in the Frog Puzzle the ability to assess the probability of certain assumptions holding true, such as the frogs mingled thoroughly, no new frogs were born, and the bands did not increase the death rate of the banded frogs).

H5 or Correlational Thinking In spite of some exceptions, the individual is able to recognize correlations between factors by comparing the number of confirming to disconfirming cases (e.g., to establish a link between hair color with eye color, the number of blue-eyed blondes and brown-eyed brunettes is compared to the number of brown-eyed blondes and blue-eyed brunettes).

Hypothetical thinking patterns, taken together, enable people to accept hypothesized statements (assumptions) as the starting point for reasoning about a situation and to reason hypothetico-deductively. In other words, they are able to imagine possibilities, deduce the consequences of those possibilities, then observe whether or not those consequences do in fact occur (Type B thinking.)

Table 3.1 summarizes important differences between these two levels of thought. If you found yourself responding to the puzzles primarily with descriptive thinking patterns, you will need to read and study Chapters 4 and 5 ("Improving Your Scientific Thinking Skills, Parts 1 and 2") carefully.

TABLE 3.1 CHARACTERISTICS OF DESCRIPTIVE AND HYPOTHETICAL THOUGHT

DESCRIPTIVE THOUGHT	HYPOTHETICAL THOUGHT
Needs reference to familiar actions, objects, and observable properties.	Can reason with second-order relationships, hypothetical properties, postulates, and theories; uses symbols to express ideas.
Thinking is initiated with observations.	Thinking can be initiated with imagined (nonobservable) possibilities.
Uses thinking patterns D1–D3. Patterns H1–H5 are either not used or used only partially, unsystematically, and only in familiar contexts.	Uses thinking patterns H1–H5 as well as D1–D3.
Needs step-by-step instructions in a lengthy procedure.	Can plan a lengthy procedure given certain overall goals and resources.
Is unconscious of his or her own thinking patterns, inconsistencies occur among various statements he or she makes, statements may be contradicted by known facts.	Is conscious and critical of his or her own thinking patterns, actively seeks checks on the validity of his or her conclusions by appealing to other known information.

Let's now turn to an example from biology to see how the hypothetical thinking patterns work together to answer a question about nature. The question is about the migration of silver salmon from the Pacific Northwest.

HOMING IN SILVER SALMON

Raising a Question

Silver salmon are found in the cool, quiet headwaters of freshwater streams in the Pacific Northwest. Young salmon swim downstream to the Pacific Ocean, where they grow to full size and mature sexually. They then return to the freshwater streams and swim upstream, often jumping incredible heights up waterfalls to lay their eggs in the headwaters before they die. By tagging young salmon, biologists were able to discover that mature salmon actually migrate to precisely the same headwaters in which they were hatched some years earlier. This discovery raised a very interesting question: How do the salmon find their way to the streams of their birth?

Creating Hypotheses and Combinatorial Thinking

We can propose a number of possible answers to this causal question. For instance, we know that people can navigate by sight. Perhaps the salmon do so as well. That is, they may remember certain objects, such as large rocks, that they saw when they swam downstream on their way to the ocean. Studies of migratory animals may also suggest answers. For example, it was discovered that migratory eels, who, like salmon, migrate from freshwater to saltwater and return, are enormously sensitive to dissolved minerals in water. Evidence suggests that a single eel can detect the presence of a mineral in water at concentrations as low as two to three molecules per liter. Perhaps the salmon are also sensitive to smell and smell their way back. They may swim a short distance into various streams until they find the one that smells right. In other words, they may be able to use their noses to detect certain chemicals specific to their home stream and follow a chemical path home.

Evidence also suggests that homing pigeons are able to navigate using the Earth's magnetic field. Pigeons wearing little magnets on their backs (to disrupt the magnetic field near the bird) were not as successful at finding their way home at night as a similar group of pigeons who wore little nonmagnetic metal bars on their backs. Perhaps the salmon are also sensitive to the magnetic field and this is how they find their home stream. Thus the process of abduction (borrowing ideas from past experiences and using them as possible explanations in this new context) gives us three tentative hypotheses:

1. Salmon use sight to find their way home (i.e., they see certain landmarks recalled from their previous trip to the ocean).

2. Salmon smell certain chemicals in the water that are specific to their home stream.

3. Salmon are sensitive to the Earth's magnetic field and use it to navigate.

It should be pointed out that other possibilities remain. Indeed, none of these three may be the "correct" answer. Or, perhaps, salmon are able to use all three or perhaps two of these three. Psychologists refer to this process of thinking of all possible combinations of hypotheses as **combinatorial thinking**. Combinatorial thinking was needed to generate all 15 possible diets of algae in the Algae Puzzle. In this case, combinatorial thinking gives us these possibilities:

1. None of the three hypotheses is correct.
2. Only the sight hypothesis is correct.
3. Only the smell hypothesis is correct.
4. Only the magnetic field hypothesis is correct.
5. Both the sight and smell hypotheses are correct.
6. Both the sight and magnetic field hypotheses are correct.
7. Both the smell and magnetic field hypotheses are correct.
8. The sight, smell, and magnetic field hypotheses are all correct.
9. One or more of the identified hypotheses in combination with one or more of the as yet to be identified hypotheses are correct.

Generating Predictions

Once the process of abduction and combinatorial thinking have provided a number of likely possible explanations, the next task is to put one or more of them to the test. For example, biologists tested the sight hypothesis by capturing and blindfolding a group of salmon and comparing their ability to locate their home stream with the ability of a similar group of nonblindfolded salmon. The thinking pattern used to test the sight hypothesis was as follows:

Hypothesis: *If*...silver salmon locate their homestream by sight (the non-observable cause)...

Experiment: *and*...a group of blindfolded (nonsighted) salmon is compared to a group of nonblindfolded (sighted) salmon, all other things being equal...

Prediction: *then*...the blindfolded salmon should not be able to locate their home stream and the nonblindfolded salmon should be able to locate their home stream (the observable result).

This pattern of hypothetico-deductive thinking is involved in all hypothesis testing.

Identifying and Controlling Variables

Notice, however, that the prediction in the previous argument follows only to the extent that the experiment is conducted in a manner that allows only the fish's sight to vary. In other words, some fish could see and others could not, but other than this, the fish were identical. In this case the fish's ability to see is called the **independent variable.** It is called an independent variable because (a) it varies from one group to another, and (b) this ability is independent of other abilities such as the ability to smell, to hear, to taste, and so forth. All values of the other possible independent variables, such as those just mentioned, must be the same to the extent possible. In other words, the experiment should be controlled. (More will be said about this in Chapter 5.)

Suppose, for example, that we do, in fact, find a difference in the ability of the two groups of fish to find their homestream (the dependent variable). This is the result predicted by the sight hypothesis; therefore, we could conclude that the difference is due to sight and not to some other variable. However, during the experiment, the blindfolded salmon may have been hindered, not by lack of sight but by an inability to swim with blindfolds. Or perhaps simply taking the fish out of the water to fit them with blindfolds shocked the fish and disrupted their ability to swim. At any rate, the experimenter must try to avoid these problems as much as possible. But because one can never be certain that all such problems have been eliminated, we have to interpret the result of all experiments with caution. Consequently, as mentioned in Chapter 2, positive results can never be interpreted as "proof" of the correctness of a hypothesis. Positive results merely allow you to conclude that the hypothesis has been supported.

On the other hand, suppose both groups of salmon are equally successful at finding their home streams. Now we can be reasonably sure that the salmon use some other means of navigation; that is, the sight hypothesis is not supported. However, again we can not conclude that the sight hypothesis has been disproved. Logically the hypothesis has been disproved, but the real world is not that simple. Again, other unknown independent variables may be operating. For example, perhaps the salmon could see under the blindfolds. This is precisely how blindfolded magicians are able to see. Or perhaps the blindfolds were not thick enough to block out all light. Or perhaps the blindfolds were indeed effective and the salmon do indeed use sight when they can, but when their sight is blocked, they use some other sense (e.g., smell) to navigate.

Drawing Conclusions

In short, the thinking involved in testing hypotheses does indeed follow a hypothetico-deductive pattern, but it also involves thinking that can be referred to as the identification and control of variables. The identification and control of variables is an absolutely crucial thinking pattern, but it is limited in that one can never be certain that all of the potentially relevant variables have been identified and/or controlled. Thus, all scientific conclusions, whether supportive or not supportive of a particular hypothesis, must remain tentative to some extent. Scientific arguments and evidence can most certainly be convincing beyond a reasonable doubt, but to the critical mind they can never be convincing beyond any doubt. In other words, as mentioned previously, hypotheses are neither proven nor disproven.

Probabilistic and Correlational Thinking

Let us return to the salmon example to discuss two additional patterns of scientific thinking. To actually test the sight hypothesis, biologists captured and tagged salmon that had just returned to spawn in one of two streams near Seattle, Washington. The streams were the Issaquah and East Fork (see Figure 3.3). The biologists then blindfolded a portion of the salmon and returned them about three-quarters of a mile below the point where the streams join. The fish were then recaptured in traps about a mile above the junction of the two streams as they made their way back up to the headwaters. Interestingly, the blindfolded salmon were as successful as the nonblindfolded salmon at finding their homestreams. Thus, the sight hypothesis was not supported.

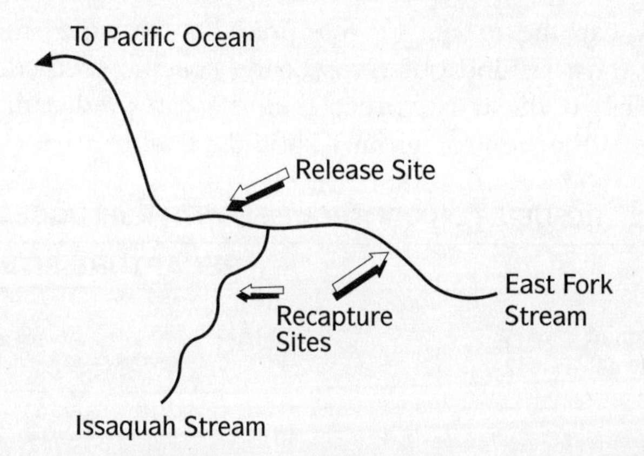

Figure 3.3 A map of Issaquah and East Fork streams, showing salmon release and recapture sites.

The smell hypothesis was tested in a similar manner. Let us consider this test, including a look at the actual numbers of fish involved. First, 302 salmon were captured from the headwaters of the two streams. The biologists then tagged the fish and divided the East Fork fish into two groups. They inserted cotton plugs coated with petroleum jelly in the noses of one of the groups to block their ability to smell. These fish were referred to as the *experimental group.* They left the noses of the other group—the *control group*—unplugged. They then divided the Issaquah fish into two groups and plugged the noses of one group just as they had done with the East Fork fish.

Finally, all the fish were taken to the release site. The fish were then recaptured in traps above the junction of the two streams (marked recapture sites on map) as they returned upstream.

Not all of the 302 tagged fish were recaptured. Some of the fish either swam downstream or swam upstream but avoided the traps. Of the 153 tagged fish with unplugged noses, 45% were recaptured while 49% of the 149 tagged fish with unplugged noses were recaptured. Additional results of the experiment are shown in Tables 3.2 and 3.3. Table 3.2 shows the results for the fish with unplugged noses (the control group). As you can see, 46 of the Issaquah fish were recaptured in the Issaquah and none were recaptured in the East Fork. Eight of the East Fork fish were recaptured in the Issaquah and 19 in the East Fork.

Table 3.3 shows the results for the fish with the plugged noses (the experimental group). Thirty-nine of the Issaquah fish were recaptured in the Issaquah and 12 in the East Fork. Sixteen of the East Fork fish were recaptured in the Issaquah and three in the East Fork.

How can these data be interpreted to test the smell hypothesis? First we must remind ourselves of the specific prediction involved. If the smell hypothesis is correct, then we can predict that the fish that can smell (the control group) should be able to locate the correct

TABLE 3.2 RESULTS FOR CONTROL FISH WITH UNPLUGGED NOSES

	RECAPTURE SITE	
CAPTURE SITE	**ISSAQUAH**	**EAST FORK**
Issaquah	46	0
East Fork	8	19

TABLE 3.3 RESULTS FOR EXPERIMENTAL FISH WITH PLUGGED NOSES

	RECAPTURE SITE	
CAPTURE SITE	ISSAQUAH	EAST FORK
Issaquah	39	12
East Fork	16	3

stream while those that cannot smell, (the experimental group) should not be able to locate the correct stream. Notice in Table 3.2 that 46 of the 46 Issaquah fish were recaptured in the Issaquah stream. Therefore, 100% of these fish were successful, as predicted by the hypothesis. But only 19 (70%) of the 27 East Fork fish found the correct stream. Eight (30%) of the 27 East Fork fish were not successful. Clearly the result from these eight fish contradict the hypothesis. Why do you suppose the control Issaquah fish were so much more successful than the control East Fork fish?

Let's now turn to the results in Table 3.3 for the experimental fish with plugged noses. Here 39 (77%) of the 51 of the Issaquah fish were successful in finding the correct stream. The East Fork fish were not nearly as successful, as only 3 (16%) of the 19 of them were recaptured in the East Fork stream. Thus, in both control and experimental groups, the Issaquah fish were, for some reason, more successful than the East Fork fish at reaching their home stream. Perhaps the Issaquah stream is wider that the East Fork at the point where they join. Whatever the reason, the overall results reveal that 46 + 19 = 65 (89%) of the 73 control group fish were successful at reaching their homestream, whereas only 39 + 3 = 42 (60%) of the 70 experimental group fish were successful. The predicted percentages based upon the smell hypothesis would be 100% for the control fish and 50% for the experimental fish. The prediction of 50% assumes that the fish with plugged noses will turn in to one of the two streams at random; therefore 50 of 100 or 50% of those turns will be successful due to chance alone.

We see that the actual experimental results (89% versus 60%) are not precisely those predicted by the smell hypothesis (100% versus 50%). But consider this: Suppose the fish's ability to smell contributes nothing to its ability to navigate. Then the percentages for each group should be the same. Therefore, the question we need to ask is this: Is the 89% success rate of the control fish significantly better than the 60% success rate of the experimental fish? Of course, we would need a statistical analysis to determine the probability of this difference occurring due to chance, but you should have the sense that given the number of fish involved, 89% is probably significantly higher than 60%. In other words, the difference is probably due to the fish's ability to smell and not due to chance. Therefore, we can conclude that the smell hypothesis has been supported.

The thinking pattern we have used to come up with such probability statements as 19 of 27, 3 of 19, or 65 of 73 is called probabilistic thinking. Further analysis of the data also required us to compare these probabilities to determine whether or not they appear to depart significantly from chance. So, for example, when we concluded that 89% was in fact a significantly greater probability than 60%, we were using what psychologists refer to as correlational thinking. The use of a correlational thinking pattern allowed us to conclude that a correlation most likely does exist between the fish's ability to smell (the independent variable) and their ability to locate their home stream (the dependent variable).

Probabilistic and correlational thinking can also be used to analyze data gathered from nonexperimental situations. For example, a 25-year study of 1,969 men found a direct correlation between the amount of cholesterol in the diet and risk of heart attack. The study reported that men who consumed more than 500 milligrams of cholesterol per day had a 22% greater probability of dying of a heart attack than men whose cholesterol intake was substantially below the 500 milligram level.

CREATIVE AND CRITICAL THINKING SKILLS

Although this is by no means a complete discussion of the creative and critical thinking skills employed by biologists, it should provide a framework for understanding important thinking patterns used in creating and testing hypotheses. Science is essentially a process of accurately describing nature and attempting to create and test hypothetical systems that serve to explain natural phenomena. Thinking skills and patterns essential to the accurate description and explanation of events in nature can be divided into the following seven categories:

1. Skill in accurately describing nature
2. Skill in sensing and stating causal questions about nature
3. Skill in recognizing, creating, and stating alternative hypotheses and theories
4. Skill in generating logical predictions
5. Skill in planning and conducting controlled experiments to test hypotheses
6. Skill in collecting, organizing, and analyzing relevant experimental and correlational data
7. Skill in drawing and applying reasonable conclusions*

Some of these skills are creative, while others are critical. Still others involve both creative and critical aspects of scientific thinking. A skill is defined as "the ability to do something well." Skilled performance includes knowing what to do, when to do it, and how to do it. In other words, being skilled at something involves knowing a set of procedures, knowing when to apply them, and being good at performing them.

These thinking skills operate in the mind of creative and critical thinkers as they learn about the world. When they are testing hypotheses, their thinking follows a hypothetico-deductive pattern; it begins not with what is observed (the real) but with what is imagined (the possible). The hypothetico-deductive thinking pattern includes key steps and the key words *if, and, then, therefore.* For example, *if* salmon navigate by sight, *and* some salmon are blindfolded, *then* they should not be able to navigate. But blindfolded salmon were able to navigate, *therefore* they must have another way to navigate. The thinking patterns and skills of the hypothetico-deductive thinker are, in essence, learning tools essential for success and even for survival. One who possesses these skills has "learned how to learn." Indeed, if you improve your use of these creative and critical thinking skills, you become more "intelligent" and have "learned how to learn."

*After Mary Alice Burmester, "Behavior involved in critical aspects of scientific thinking." *Science Education,* 36 (5), 259–263, 1952.

QUESTIONS

Answer each of the following questions and then identify some of the thinking patterns involved. How might a descriptive thinker respond to each question? How might a hypothetical thinker respond?

1. Figure 3.4 depicts a wide and a narrow cylinder. The cylinders have equally spaced marks on them. Water is poured into the wide cylinder up to the fourth mark (see A). This water rises to the sixth mark when poured into the narrow cylinder (see B). Water is now poured into the wide cylinder up to the sixth mark. How high would this water rise if it were poured into the empty narrow cylinder?

   ___

   Show (or explain) how you arrived at your answer.

   ___

   ___

   ___

   ___

2. Water is now poured into the narrow cylinder (described in question 1) up to the eleventh mark. How high would this water rise if poured into the empty wide cylinder?

   ___

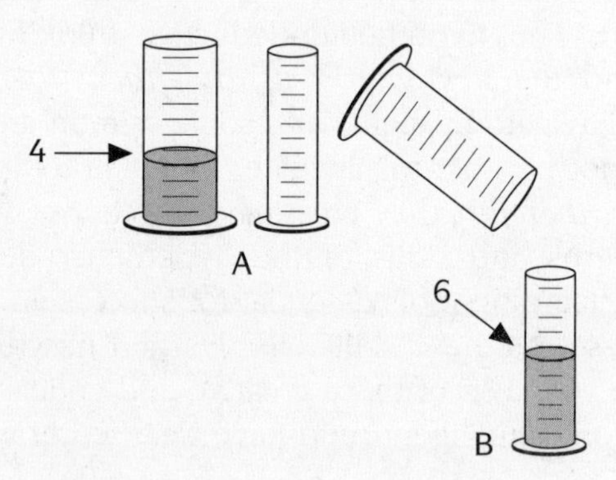

Figure 3.4 Thinking about cylinder size, shape, and volume.

Show (or explain) how you arrived at your answer.

____

____

____

____

3. Figure 3.5 depicts three strings hanging from a bar. The three strings have metal weights attached to their ends. String 1 and string 3 are the same length. String 2 is shorter. A 10-unit weight is attached to the end of string 1. A 10-unit weight is also attached to the end of string 2. A 5-unit weight is attached to the end of string 3. The strings (and attached weights) can be swung back and forth, and the time it takes to make a swing can be timed. Suppose you wanted to find out whether length of string has an effect on the time it takes to swing back and forth. Which strings would you use to find out?

   ____

   Explain why you chose those strings.

   ____

   ____

   ____

   ____

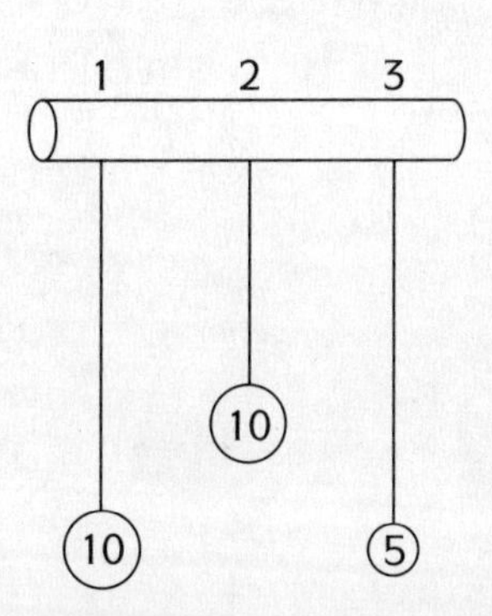

Figure 3.5 Thinking about string length and swing time.

4. Suppose you wanted to find out whether the amount of weight attached to the end of a string has an effect on the time it takes for a string to swing back and forth. Which of the strings in question 3 would you use to find out?

___

Explain why you chose those strings.

___

___

___

___

5. Twenty flies are placed in each of four glass tubes. The tubes are sealed. Tubes 1 and 2 are partially covered with black paper; tubes 3 and 4 are not covered. The tubes are suspended in midair by strings as shown in Figure 3.6. Then they are exposed to red light for five minutes. The number of flies in the uncovered part of each tube is shown in Figure 3.6. This experiment shows that flies respond to (go to or away from)

___

a. red light but not to gravity.

b. gravity but not to red light.

c. both red light and gravity.

d. neither red light nor gravity.

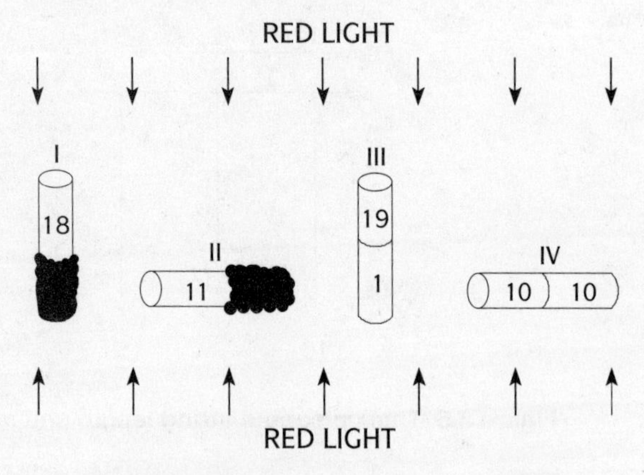

Figure 3.6 Fly response to red light.

Explain your selection.

____

____

____

____

6. In a second experiment, blue light was used instead of red. The results are shown in Figure 3.7. These data show that flies respond to (go to or away from)

____

a. blue light but not to gravity.

b. gravity but not to blue light.

c. both blue light and gravity.

d. neither blue light nor gravity.

Explain your selection.

____

____

____

____

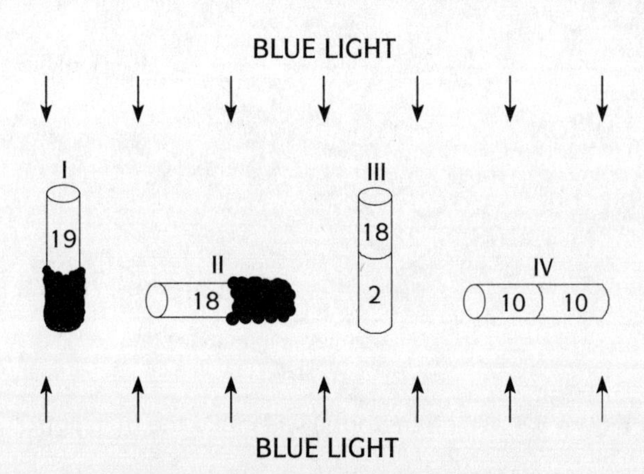

Figure 3.7 Fly response to blue light.

7. Three red, four yellow, and five blue square pieces of wood are put into a cloth bag. Four red, two yellow, and three blue round pieces are also put into the bag. All the pieces are then mixed about. Suppose someone reaches into the bag (without looking and without feeling for a particular shape) and pulls out one piece. What are the chances that the piece is a red or blue circle?

   ____

   Show (or explain) how you arrived at your answer.

   ____

   ____

   ____

   ____

8. Figure 3.8 shows a box with four buttons numbered 1, 2, 3, and 4 and a light bulb. The bulb will light when the correct button, or combination of buttons, are pushed together. Your problem is to figure out which button, or buttons, must be pushed all at the same time to make the bulb light. Make a list of all the buttons and combinations of buttons you would push to figure out how to make the bulb light.

   ____

   ____

   ____

   ____

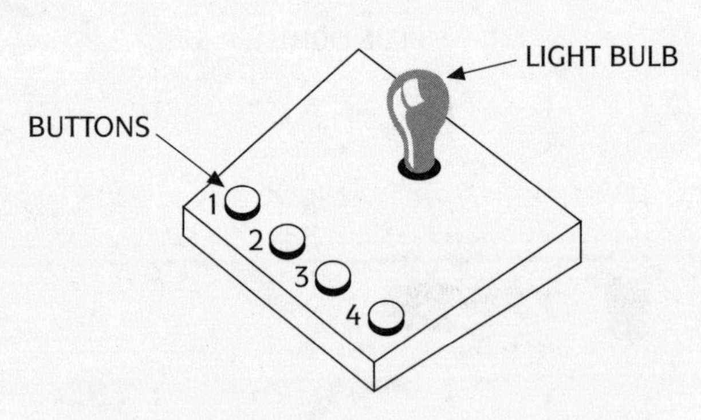

Figure 3.8 Which button(s) will make the bulb light?

9. Figure 3.9 shows fish that were caught by a fisherman. The fisherman noticed that some of the fish were big and some were small. Also, some had wide stripes and others had narrow stripes. This made the fisherman wonder if there was a relation between the size of the fish and the width of their stripes.

   Do you think there is a relation between the size of the fish and the width of their stripes?

   ____

   a. Yes

   b. No

   Explain your answer.

   ____

   ____

   ____

   ____

Figure 3.9 Is there a connection between fish size and stripe width?

TERMS TO KNOW

- combinatorial thinking
- correlational thinking
- dependent variable
- independent variable
- probabilistic thinking
- proportional thinking
- thinking pattern

CHAPTER 4

IMPROVING YOUR SCIENTIFIC THINKING SKILLS: Part I

GETTING FOCUSED

- *Identify the basic elements of scientific thinking embedded in a story.*
- *Develop skill in using probabilistic thinking.*
- *Develop skill in using correlational thinking.*

This chapter and the next should help you become a better scientific thinker. Let us start by trying to identify the general elements of scientific thinking embedded in a story about a famous horse named Clever Hans. We will then turn to a discussion of probabilistic and correlational thinking and work through several problems to help you develop skill in using these very important thinking patterns. Other thinking patterns will be the subject of Chapter 5.

THE CASE OF CLEVER HANS

Clever Hans by all accounts was a very clever horse. According to German newspapers around the turn of the century, he could identify musical intervals, understand German, and was quite good at math. When his owner, Herr Wilhelm von Osten, asked him to add numbers, he tapped out the answer with his hoof. For other questions he gestured with his head toward the appropriate pictures or objects.

Understandably, many people were skeptical about this, so a group of "experts," including two zoologists, a psychologist, a horse trainer, and a circus manager, were brought in to investigate. They watched closely as Herr von Osten asked question after question to which Hans replied with near-perfect accuracy. Hans was even able to reply correctly to questions posed by strangers. If the question called for the square root of 16, Hans would confidently tap four times with his hoof. The experts were unable to discover any tricks, thus were forced to conclude that Hans was a very clever horse indeed.

Do you agree with the experts? If not, how could Hans have correctly answered the questions? After reading the experts' report, a young psychologist named Oskar Pfunget proposed a different explanation. Suppose Hans was not able to think out the answers at all. But suppose instead he was able to monitor subtle changes in the questioner's facial expressions, posture, and breathing that would occur when Hans arrived at the correct answer. Perhaps these cues could tell Hans when to stop tapping or moving his head.

How could this explanation be tested? Oskar decided he needed to have some questioners who did not know the correct answers. If the questioner did not know the answers, then his expressions, and so forth, could not clue Hans, so Hans's success rate should drop considerably. Sure enough, when the interrogator knew the answer, Hans succeeded on 9 out of 10 problems. But when the interrogator was ignorant, Hans score dropped to just 1 out of 10. Hans had not learned math, music, or German after all. Instead he had learned how to read people's faces and their body language! In one sense Hans was not smart at all, but in another sense he was extremely perceptive.

Now let's see if you can identify the key elements of the scientific method from the story. Use Figure 4.1 on page 82 to do so by filling in the boxes with the correct questions or statements. You can compare you answers with mine (shown in Figure 4.2).

As the story of Clever Hans shows, scientific thinking includes raising questions about causes, generating tentative explanations (hypotheses), imagining ways to test the tentative explanations, deducing predicted results, organizing and analyzing data, and drawing reasonable conclusions. People who have acquired the habit of thinking scientifically seldom jump to hasty conclusions. Instead, the path to knowledge is a long one that passes from speculation about possible causes, to the explicit statement of alternative hypotheses, to the testing of working hypotheses, to the consideration of supported or not supported hypotheses, and finally to the formulation of beliefs about causes based upon a careful consideration of the alternatives, the evidence, and other related sets of beliefs. In the final analysis, the scientific thinker believes some explanation is true or false, not merely because some authority says so, or because it may "seem to make sense," but because the alternatives and the evidence have been given fair and impartial consideration. If any of these elements of scientific thinking are missing, the proper conclusion for the scientific thinker is to say, "I do not know."

Of course, there is more to scientific thinking than identifying hypotheses, experiments, predictions, results, and conclusions. The whole business can get quite complicated and involved. As you learned in Chapter 3, scientific thinking also involves more specific thinking patterns, such as the identification and control of variables, combinatorial thinking, and probabilistic thinking. The remainder of this chapter will provide more details about these more specific thinking patterns and some problems you can solve to help you acquire these patterns. We start with probabilistic thinking.

PROBABILISTIC THINKING

What are your chances of winning a $5,000 jackpot? What are your chances of being struck by an automobile while crossing the street? What are your chances of living to be 100 years old? Do frogs with spots have a better chance of survival than frogs with no spots? Do

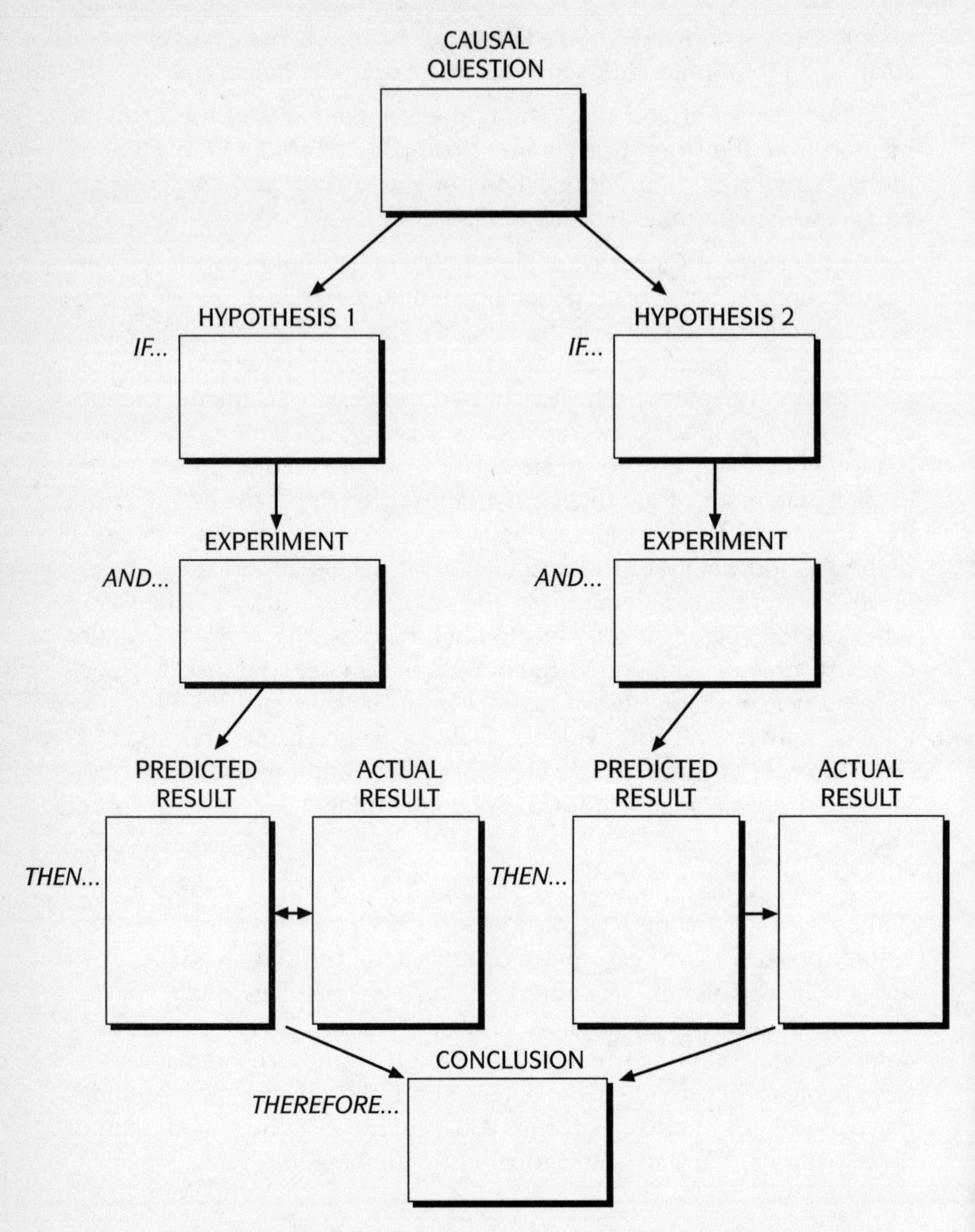

Figure 4.1 Fill in the boxes with the appropriate questions and/or statements from the story about Clever Hans.

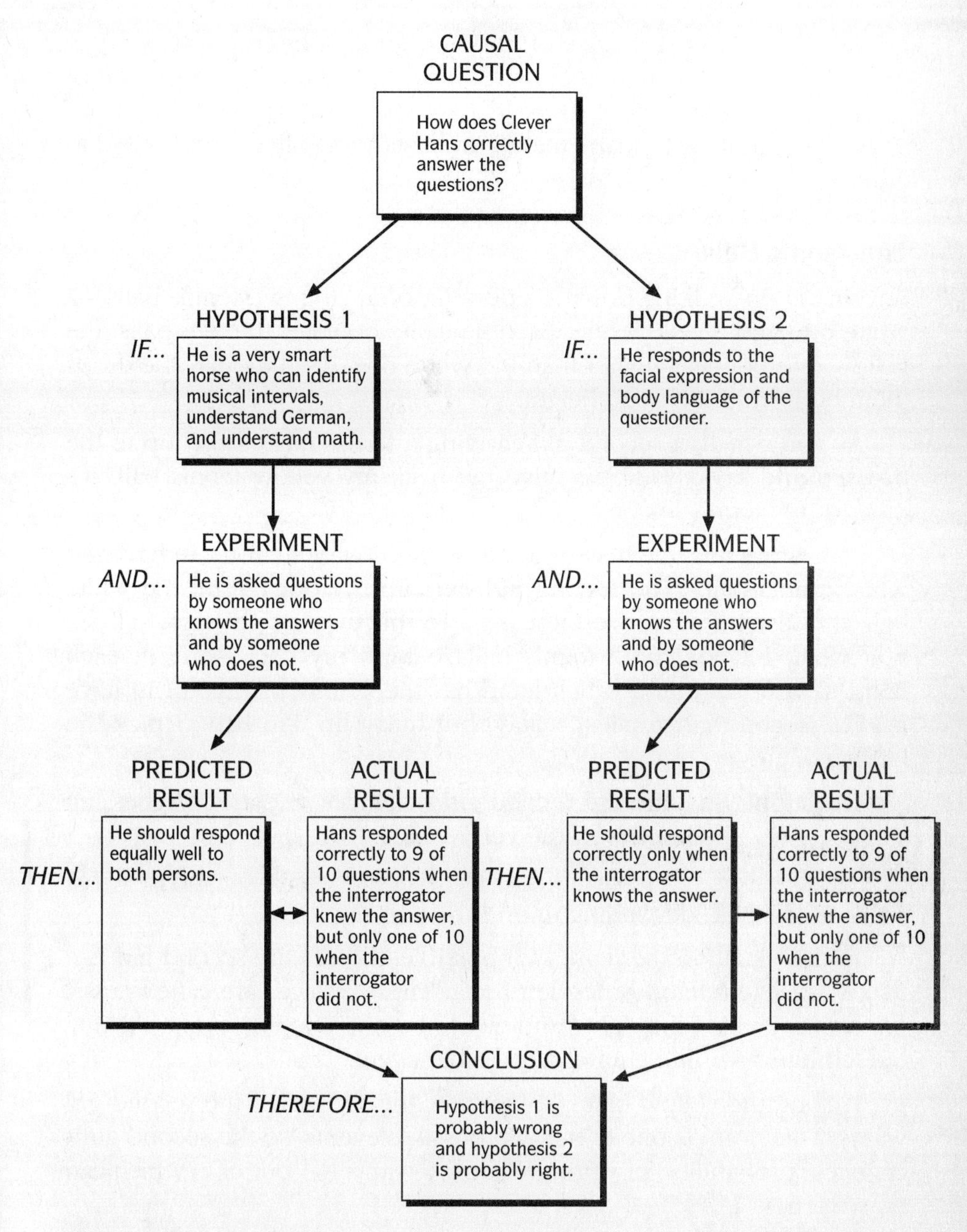

Figure 4.2 The major elements of scientific thinking as identified in the story about Clever Hans.

farmers have a better chance of raising a good corn crop if they spray with insecticide? The notion of probability is embedded in every part of our lives as well as that of every living organism. Understanding probability is, therefore, fundamental to understanding life and the world in which we live.

The Tennis Balls

Imagine a girl walking down a sidewalk, bouncing two tennis balls—a white one and a yellow one. Each time she bounces the two balls, the yellow one bounces higher than the white one, even though she drops them both from the same height.

As you watch, a second girl carrying a tennis racket runs up to the first girl and says, "Hey, Sis, give me back my yellow tennis ball. It's time for my tennis lesson."

At hearing this, the first girl replies, "No. I want it. You can have the white tennis ball." The second girl exclaims, "I don't want the white ball. It doesn't bounce worth beans." To this the first girl says, "I'll tell you what. I'll hold both tennis balls behind my back—one in each hand. If you can guess which hand has the yellow ball, you can have it." The second girl replies, "Okay, but hurry up. I'm late...I pick the left hand."

Question: What are the second girl's chances of correctly guessing the hand with the yellow ball?

The answer, of course, is that the chances are 1 out of 2, 50-50, 50%, or one-half, depending upon how you say it.

In this situation there are two possible choices the second girl can make—the right hand or the left hand. These choices are called *possible events*. Picking the right hand would be one possible event. Picking the left hand would be another possible event.

Picking the yellow ball is, of course, the event that the second girl wants. This event is one of the two possible events, so the second girl's chances (probability) of getting what she wants is 1 out of 2, which can be written as 1/2.

1 (event of picking yellow)/1 (event of picking yellow)+1 (event of picking white)= 1/2

The other answers, such as 50%, 1 to 1, or 50-50, also describe the chances (i.e., the probability). Fifty percent means that if you tried guessing 100 times, you would guess correctly about one-half the time or 50 times=50 per 100=50%. One to 1 means that for every 1 event that is correct (picking the hand with the yellow ball), there is 1 event that is not correct (picking the hand with the white ball). Fifty-fifty means that if you guessed 100 times, one-half the time or 50 times you would pick the yellow ball, and one-half the time or 50 times you would pick the white ball.

A Sack of Beans

The probability (chances) of the second girl picking the yellow tennis ball was 1 out of 2 or 50%. But what does this 50% mean anyway? Does it mean that if the second girl actually guessed which hand held the yellow ball 100 times, she would guess correctly exactly 50 times? Let's see just what this does mean.

Find a paper sack and some brown and white beans. Put 50 brown beans and 50 white beans in the sack.

Now close the sack and shake it up. Without looking into the sack, pick out 10 beans at random. How many of the beans are brown? Plot this number on the graph in Figure 4.3 above the place for trial number 1. Now put the 10 beans back into the sack and shake it again.

Pick out another 10 beans at random. How many of these are brown? Plot this number on the graph in Figure 4.3 above the place for trial number 2. Again put the 10 beans back into the sack and shake it.

Repeat this procedure eight more times.

Now look at your graph.

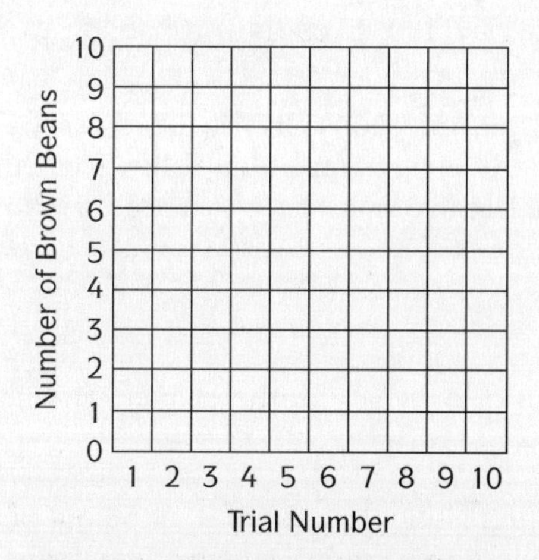

Figure 4.3 Determining probability.

Does it appear as though "on the average" 5 out of every 10 beans are brown? __________

Is this the same as saying "on the average" 1 out of every 2 beans are brown? __________

Does 5/10 = 1/2 = 50/100? __________

Why didn't you get 5 brown beans and 5 white beans on each trial?

What is the total number of brown beans picked out for all ten trials? Why is this number likely to be close to 50 but unlikely to be exactly 50? __________

In this activity, the 100 beans can be called a *population*. You actually obtained 10 different *samples* of the population. Even if you did not know the number of brown beans in the population, you could use each sample to estimate (guess) characteristics of the population. In this case, we could attempt to estimate the portion of the population that is brown. That portion was 50 out of 100, or 5 out of 10, or 1 out of 2. Therefore, in this population, the probability of being brown is 50%. Notice, however, that each sample provided an estimate of the population. All sample estimates contain some error. The idea is to make the error as small as possible by getting as large a sample as possible without going to too much trouble or expense.

In short, the probability of any desired event(s) is given as follows:

$$\text{Probability} = \frac{\text{Number of desired event(s)}}{\text{Total number of possible event}}$$

Multiplying this by 100 gives the probability expressed as a percentage.

Try the following questions. If you need some help in directing your thinking, read the paragraph that follows each of the first three questions. For the last five questions, you are on your own.

Questions

1. If you flip a coin, what are the chances it will come up heads?

 ____

 Again, the probability is 1 out of 2, or 50 out of 100, or 50%. The event "coming up heads" is what we want. There are two equally possible events—"coming up heads" or "coming up tails"—so the answer is 1 event out of 2 possible events.

2. When you roll a die, what is the probability you will roll a 6?

   ____

   The event you want is "rolling a 6." What are the possible events? "Rolling a 1," "rolling a 2." "rolling a 3," "rolling a 4," "rolling a 5," "rolling a 6"—a total of 6 different possible events. The chances of rolling a 6, then, are 1 out of 6 or 16.6% (1/6 = 1 ÷ 6 = .166, .166 × 100 = 16.6%).

3. Suppose 5 diamonds and 20 pieces of cut glass are put into a sack. What are your chances of reaching into the sack and pulling out a diamond on the first try?

   ____

   The event you want is "picking a diamond." There are 5 ways of doing this since you could pick any one of 5 diamonds. There are 25 possible events since there are 25 pieces in all, so your chances of "picking a diamond" are 5 out of 25 or 20% (5/25 = 5 ÷ 25 = .20, .20 × 100 = 20%).

4. A farmer collects 25 field mice from a corner of one of his fields and notes that 20 of the mice are spotted and 5 are all white. How many mice are in the farmer's sample? What fraction of the sample of mice is spotted? __________

   What percentage of the sample is spotted? __________

   Suppose he captures one more mouse. What is the probability that the mouse will be spotted? __________

5. In a small town in northern Arizona, 75 people were given the swine flu vaccination last winter. Five of those 75 people developed muscular paralysis within two weeks of receiving the vaccine. What portion of the people developed paralysis?

   ____

   Suppose you were to go to that town to receive the vaccine. What would you estimate your chances are of developing paralysis?

   ____

6. On the first day of class in a college biology class for nursing majors, 23 out of the 28 students enrolled were females. What percentage of the class was female? __________

   On the second day of class, a new student enrolled. What were the chances that this student was a male? __________

   Does this fact mean that only about 18% of the population of students at this school are male? __________

   Explain your answer.

   ____

   ____

7. What are your chances of finding a needle in a 5-foot-tall haystack within 1 hour?

   ____

CORRELATIONAL THINKING

Does smoking cause lung cancer? Does Laetrile cure cancer? Does drunk driving cause accidents? Do fatty foods cause high blood pressure? Does increased radiation increase mutations? Do thin people live longer? Do small organisms have higher metabolic rates than large ones?

You may think you know the answers to some of these questions already. On what evidence and/or reasoning do you base your answers? The collection and analysis of the data to determine whether or not two factors are "linked" are important components of scientific investigations. Finding factors to be linked (correlated) suggests the possibility of a cause-effect relationship. Such relationships are at the very heart of scientific understanding.

Things That Go Together

Remember the girl walking down the sidewalk, bouncing the white and the yellow tennis balls? Well, her sister finally did get the yellow ball and left for her tennis lesson. As you continue to watch, the girl walks down the sidewalk until she comes to a bumpy dirt road. She walks down the dirt road and starts to bounce the white tennis ball as she goes. You notice that each time she drops the ball it hits a different part of the road. Sometimes the ball hits a soft spot, and a low bounce results.

Since the height the ball bounces changes (varies) from bounce to bounce, the "height of bounce" is called a variable. If the height of the bounce stopped at the same place each time, it would not be called a variable. It would be called a constant.

Notice that the condition of the road also varies. Sometimes it is hard and sometimes it is soft. So "hardness of the road" would also be considered a variable. Of course, if the road were paved so that it was equally hard in all spots, the "hardness of the road" would not be considered a variable. It would be a constant.

In the present example there are two variables: (1) the distance the ball bounces after it hits the road, and (2) the condition of the road. The ball either bounces high or low and the road is either hard or soft.

Since the two variables "go together" (high bounces occur when the ball hits the hard spots and low bounces occur when the ball hits the soft spots), we say that a correlation exists between the variables "height of bounce" and "hardness of road."

A brief statement that summarizes the "linked" or correlational relationship would be as follows: the harder the road the higher the bounce. Or said another way: the softer the road the lower the bounce.

Now that you have some idea of what is meant by the terms *variable* (something that changes), *constant* (something that stays the same), and *correlation* (two things that vary together), let's see how these ideas can be applied in a new situation.

The Strange Classes at Hohumburg State University

Imagine you are visiting a friend of yours who is a student at Hohumburg State University, a somewhat strange school located in Hohumburg, Virginia. I say strange because the students at Hohumburg State never choose the classes they take. They are assigned to classes based upon their hair and eye color! Also, every course has exactly 100 students enrolled. Since this all seems so strange, you decide to investigate some of the classes.

English 101 The first class you visit is English 101. Figure 4.4 shows pictures of a sample of 20 of the English 101 students. Classify the students in the figure into two groups based upon observed differences.

- What characteristic(s) did you use for your classification?

___

- How many students are in each category?

___

Since hair color varies in the English 101 students (sometimes it is blond, sometimes it is brunette) hair color is considered a variable. English 101 students have either blond or brunette hair. Eye color, however, is constant. It is always blue. Apparently, you need blue eyes to enroll in English 101.

If our sample of 20 students is typical of the entire population of students, we can extrapolate the trend in the data to include all 100 students in English 101. If we do this, about 50 out of 100 = 50% have

Figure 4.4 Students of English 101. Ten students have brown hair, ten have blond hair. All have blue eyes.

blond hair and about 50 out of 100 = 50% have brunette hair; 100 out of 100=100% would have blue eyes.

Biology 280 The next class you visit is Biology 280. Pictures of 20 Biology 280 students are shown in Figure 4.5. Classify the students into two groups based upon observed differences.

- What characteristic(s) did you use for your classification?

___

- How many students are in each category?

___

Since eye color varies in the sample of Biology 280 students (sometimes it is blue, sometimes it is brown), eye color is a variable. Notice that hair color is constant. It is always blond. Apparently, in Biology 280 you can have either blue or brown eyes but you must have blond hair.

Again, if we extrapolate from 8 students with blue eyes out of 20 students total, we get 40 out of 100. Therefore, we can say that about

Figure 4.5 Students of Biology 280. All 20 students have blond hair. Twelve students have brown eyes and eight have blue eyes.

40% of the students in Biology 280 have blue eyes. Twelve out of 20 or 60 out of 100 = 60% of the students have brown eyes.

> **Physical Education 210** A sample of pictures of 20 Physical Education 210 students are shown in Figure 4.6. Classify these students into two groups based upon observed differences.

- What characteristic(s) did you use for your classification?

  ____

- How many students are in each group?

  ____

Notice that hair color and eye color are both variables in this class. Also notice that they go together (covary). The blond-haired students all have blue eyes and the brunette students all have brown eyes. Ten out of 20 or 50 out of 100 = 50% of the students are blue-eyed blonds, and 10 out of 20 or 50 out of 100 = 50% are brown-eyed brunettes.

We say that there is a correlation between "hair color" and "eye color" in this sample just as there was a correlation between the vari-

Figure 4.6 Students of Physical Education 210. Ten students have blue eyes and blond hair and ten have brown eyes and brown hair.

ables of "height of bounce" and "hardness of the road" in the first study. Apparently, to be enrolled in Physical Education 210, you must have either blond hair and blue eyes or brunette hair and brown eyes.

American History 320 Let's try another class and see what other relationships we can find. Look at Figure 4.7, which shows pictures of 20 students from American History 320. Classify these students into two groups.

- Do hair color and eye color go together?

Again, the answer is "yes" although not in the same way as before. Now all blond-haired students have brown eyes and all the brunette-haired students have blue eyes. Nevertheless, hair color and eye color do go together in the history students.

- What percentage of the total is in each group?

Apparently, to enroll in American History 320 you must either have blond hair and brown eyes or brunette hair and blue eyes.

Figure 4.7 Students of American History 320. Ten students have blue eyes and brown hair and ten have brown eyes and blond hair.

Calculus 100 A sample of 20 pictures of the students from Calculus 100 are shown in Figure 4.8. Classify these students into groups.

- How many different categories are there?

____

- How many students are in each category?

____

- Do hair color and eye color go together in this sample?

____

The answer is yes. Hair color and eye color do go together, but there are two students who did not fit the pattern—the student with blond hair and brown eyes and the student with brunette hair and blue eyes. These students are exceptions. Nevertheless, since most of the students fit the rule (18 out of 20 or, if we extrapolate the trend, 90 out of 100 or 90%), it still looks as though hair color and eye color go to-

Figure 4.8 Students of Calculus 100. Nine students have blue eyes and blond hair, nine have brown eyes and brown hair, one has blue eyes and brown hair, and one has brown eyes and blond hair.

gether (are correlated) in Calculus 100. Eighteen students fit the rule and only two are exceptions.

Said another way, most (9 out of 10 or 90%) of the blond-haired students have blue eyes while most (9 out of 10 or 90%) of the brunette students have brown eyes, so the variables go together, most of the time anyway.

Said still another way, the probability of having blue eyes if you have blond hair is 90%. The probability of brown eyes if you have brunette hair is also 90%.

Just why we have two exceptions to the rule is not clear. Maybe the two students sneaked into class without the teacher noticing!

Philosophy 400 So far so good. Let's visit still another class. Look at Figure 4.9, which shows pictures of students from Philosophy 400.

- How many groups are there?

____

Figure 4.9 Students of Philosophy 400. Five students have blue eyes and blond hair, five have blue eyes and brown hair, five have brown eyes and blond hair, and five have brown eyes and brown hair.

- How many students are in each group?

___

- Do hair color and eye color go together (covary) in Philosophy 400?

___

In this instance, the answer is no. Here we have five students in each of the four groups. We have just as many students who are exceptions (10) as students who fit (10) the pattern. There are at least two ways to explain why the data indicate that hair color and eye color do not go together.

1. Suppose you thought that blond hair and blue eyes should be one group and brunette hair and brown eyes should be another group. You have 10 students who fit this expectation, but you also have 10 students who do not (the blond hair-brown eyes and the brunette hair-blue eyes). Because it is 10 out of 20 or 50% that fit versus 10 out of 20 or 50% that do not fit, the number that fit is equal to the number that do not fit. Apparently, no correlation exists.
2. Can we say that most of the blond hair students have blue eyes and most of the brunette hair students have brown eyes? No, this is not the case. Only 10 out of 20 or 50% of the blond-haired students have blue eyes and only 10 out of 20 or 50% of the brunette students have brown eyes. Again, it is 50% to 50%, so it looks as though no correlation exists. The probability of having blue eyes is 50% whether you have blond hair or brunette hair. The probability of having brown eyes is also 50% whether you have blond hair or brunette hair. Hair color and eye color do not seem to go together in this class.

Let's work through one more example. Then you can try a few others if you still feel somewhat unsure of the ideas presented.

The Swim Team Tryouts

Last spring 25 women showed up for tryouts for the women's swimming team. To qualify, each swimmer was to swim the 50-yard freestyle race, with the fastest 15 swimmers making the team.

The coach noticed that often the taller women were beating the shorter women in the qualifying races. This made her wonder if there might be a correlation between height of the swimmers and the time it took them to swim the 50 yards. To find out, she recorded the swimmer's heights and their respective times for the 50-yard race, as shown in Table 4.1 on page 98. Then the coach decided to plot the data on a graph.

The graph in Figure 4.10 on page 99 shows a general inverse correlation between the variables of "height" and "time." Although there are exceptions (e.g., swimmer number 6 is shorter yet faster than swimmer number 23), in general, the taller the swimmer the less the time it took her to swim the race.

The coach then divided the graph into four sections as shown in Figure 4.11 on page 100.

1. The swimmers who were tall and fast
2. Those who were tall and slow
3. Those who were short and fast
4. Those who were short and slow

The coach then put the numbers of swimmers in each of these four categories into the table shown in Figure 4.12 on page 100. Such tables are called contingency tables because they can reveal whether the variables are "contingent" upon each other, that is, whether they are correlated.

If an inverse correlation in fact exists, we would expect to find most swimmers in the tall and fast or short and slow categories.

- Is this the case?

  ____

In general, most of the tall swimmers (9 out of 12) were fast and most of the short swimmers (10 out of 13) were slow. Said another way, 19 of the swimmers were either tall and fast or short and slow, while only 6 swimmers were exceptions (tall and slow or short and fast). So the variables of height and time do go together. A correlation exists between them.

- Do you think it is fair to categorize swimmers number 12 and number 21 as exceptions?

  ____

  Explain.

  ____

  ____

TABLE 4.1 SWIM TEAM TRYOUT DATA

SWIMMER NO.	HEIGHT (IN FEET AND INCHES)	TIME (IN SECONDS)
1	5′2″	28.0
2	5′0″	29.5
3	5′0″	30.5
4	5′7″	26.5
5	5′10″	26.5
6	5′3″	25.0
7	5′8″	25.0
8	5′11″	24.0
9	6′0″	24.5
10	5′3″	29.9
11	5′1″	26.0
12	5′6″	27.0
13	4′11″	31.0
14	5′10″	28.5
15	5′11″	25.5
16	6′1″	25.5
17	6′2″	24.0
18	5′4″	30.5
19	5′2″	29.5
20	5′5″	29.0
21	5′7″	28.0
22	5′8″	30.5
23	6′0″	26.5
24	5′5″	28.0
25	5′0″	28.5

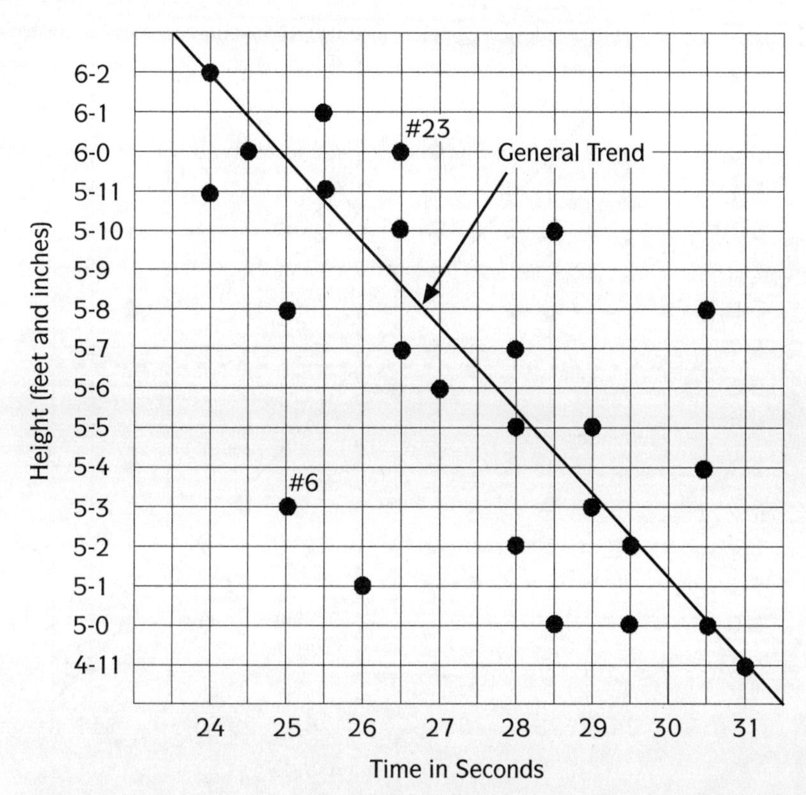

Figure 4.10 This graph shows each swimmer's height versus the time it took her to swim 50 yards. The general trend in data is indicated.

- Does height cause a swimmer to be fast?

We will discuss the issue of determining cause and effect relationships later.

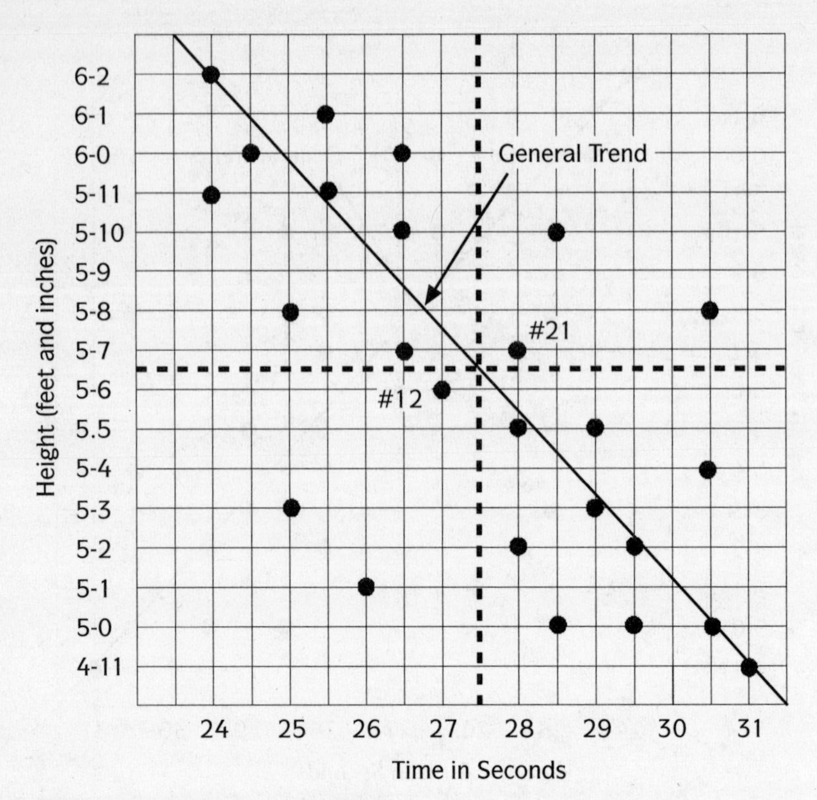

Figure 4.11 This graph is identical to the one in Figure 4.10, but it has been divided into four sections.

Height		Time: Fast	Time: Slow
	Tall	9	3
	Short	3	10

Figure 4.12 A contingency table based on Figure 4.11.

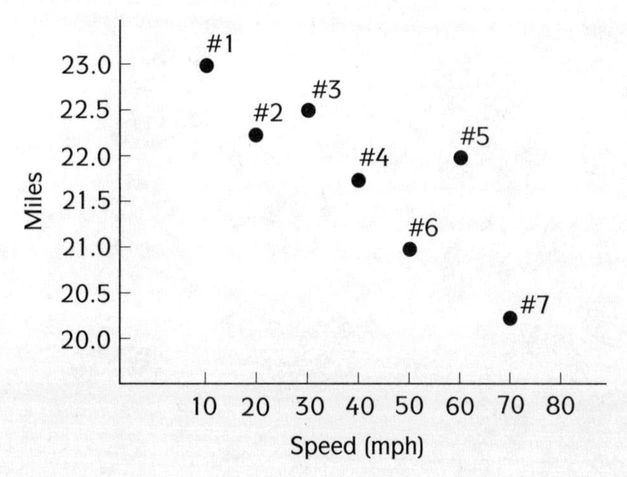

Figure 4.13 A graph showing automobile speed and miles per gallon of gasoline.

QUESTIONS

1. Figure 4.13 depicts a graph that shows the speed at which automobiles were driven and the number of miles obtained from one gallon of gasoline.

   a. Does the graph show a correlation? Explain your answer.

   ____

   ____

   ____

   b. Which points are exceptions?

   ____

   ____

   ____

   c. What could be the reason for the exceptions?

   ____

   ____

   ____

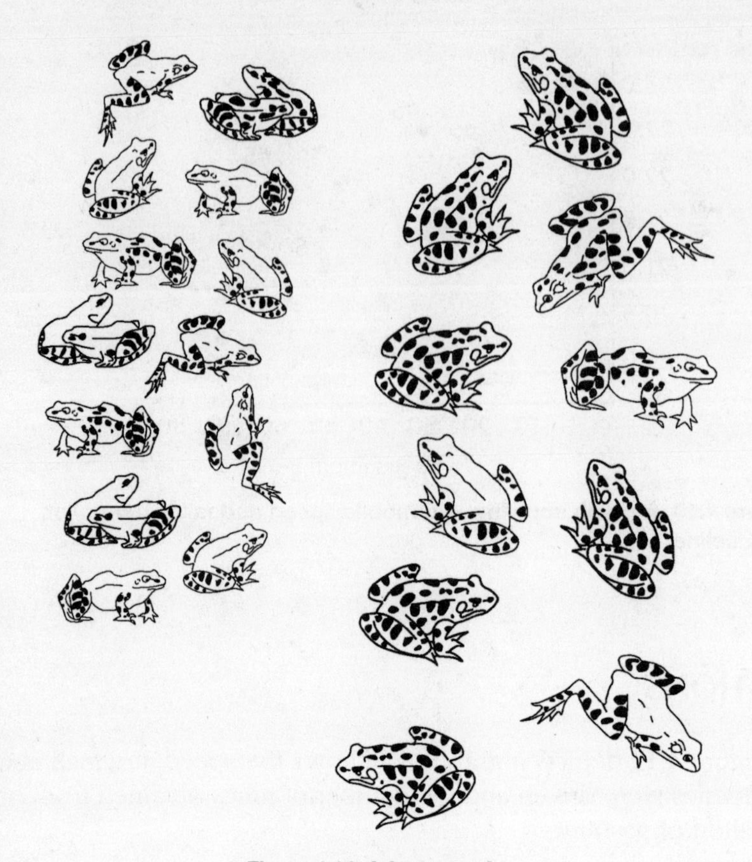

Figure 4.14 A frog sample.

2. Figure 4.14 depicts a sample of frogs collected from a pond. Does the sample show a correlation between the size of the frogs and the number of spots on their backs? Explain your answer.

____

____

____

3. The medical records from County General Hospital revealed the following data. The patient numbers were randomly selected from all of those who died at the hospital in the past 10 years. Table 4.2 indicates the cause of death and whether or not the patient smoked.

TABLE 4.2 COUNTY GENERAL HOSPITAL PATIENT DATA

		SMOKER	
PATIENT NO.	**CAUSE OF DEATH**	**YES**	**NO**
45	Lung cancer	X	
51	Internal bleeding		X
73	Lung cancer	X	
84	Natural causes		X
92	Lung cancer	X	
93	Lung cancer		X
103	Heart attack		X
104	Lung cancer	X	
121	Natural causes		X
132	Heart attack	X	
143	Stab wounds	X	
150	Lung cancer	X	
157	Breast cancer		X

Do the data show a correlation between death by lung cancer and smoking? Explain your answer.

____

____

4. Read the article below and answer the questions that follow.

New study links low pay to high blood pressure

By Cora Lation

NEW YORK—Bad news for people with low income was turned up in a recent study reported in the New York Journal of Medicine.

The study was conducted by a team of medical researchers on 50,000 people with varying degrees of increased blood pressure.

According to research team leader Dr. Mega Bucks, "The less people earn, the higher their blood pressure. The link is so strong that we can take someone's blood pressure and almost tell how much they earn."

Dr. Bucks acknowledged that, although he was not a subject in the study, his blood pressure was extremely low for his age group.

Bucks said that the results of the study are particularly disturbing because it says that people with the most severe problems are the ones least able to afford treatment.

Statistics reveal that when left untreated, one-third of people with very high blood pressure will develop health problems within a year. Yet Dr. Bucks's survey results showed that more than two-thirds of such people are not on medication.

a. What two variables are correlated according to the article?

____

____

b. Name two values for each valuable.

____

____

c. Is this a direct or inverse correlation? Explain.

____

____

d. Does this author argue that one of the variables is a cause? Explain.

____

____

6. Look through the newspaper during the next few days. Read and cut out an article that discusses variables that are correlated. Bring the article to class for discussion.

CHAPTER 5

IMPROVING YOUR SCIENTIFIC THINKING SKILLS: Part II

GETTING FOCUSED

- *Distinguish results due to chance from those due to a specific cause.*
- *Develop skill in identifying and controlling variables.*
- *Develop skill in using proportional thinking.*

In this chapter, we continue our close look at scientific thinking patterns to help you improve your thinking skills. The thinking patterns we discuss involve chance relationships, causal relationships, the identification and control of variables, and proportional relationships.

CHANCE AND CAUSAL RELATIONSHIPS

I once knew a woman who had been married six times. Her first husband was killed when his automobile was struck by a truck. Her second husband was struck down in his prime by a sudden and unexpected heart attack. Her third husband drowned while swimming in the ocean. Her fourth husband met his untimely demise due to an overdose of sleeping pills. Her fifth husband contracted a rare and exotic disease and died while on a trip to South America. Her sixth husband slipped on a roller skate that had been left at the head of a long flight of stairs. The slip resulted in a fatal tumble to the bottom.

Question: Would you marry this woman?

Sometimes a series of events just stretches the laws of probability too far. When this occurs, we assume a hidden cause. But just what is too far? To answer this question, let's take a little closer look at the issue of probability.

Two in a Row? You Must Be Kidding!

When we last left our two girls with the tennis balls, the second girl had guessed that the yellow ball was in the first girl's left hand. Was she correct? Actually, the yellow ball was in the right hand, so she was wrong.

At seeing this, the second girl exclaimed, "Oh come on, I'm in a hurry so just give me the yellow ball." To this the first girl said, "No deal, but I will be nice and give you another try. But this time you must guess correctly twice in a row before I'll let you have the yellow ball."

What are the second girl's chances of guessing correctly twice in a row?

Obviously, the chances are less than one out of two since it is tough to guess right once, but guessing right twice in a row is even tougher. Is it twice as hard? Let's see just how tough it is.

The sequence of events we want is "picking the yellow ball on the first try" and "picking the yellow ball on the second try." How many possible sequences of events are there? By using the "tree" diagram in Figure 5.1, we can generate the possible sequences.

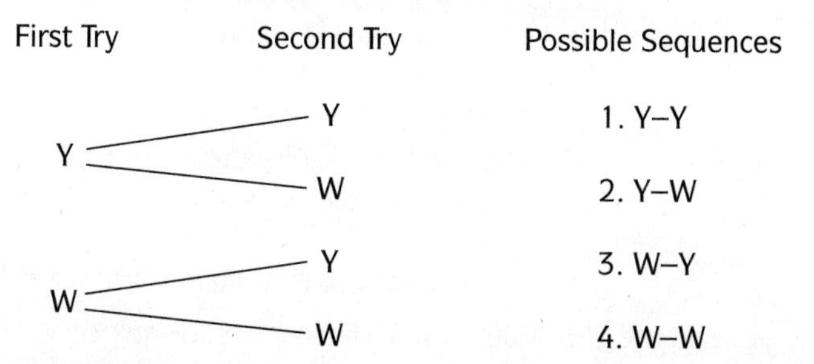

Figure 5.1 A tree diagram helps you visualize the possible sequences of events.

Figure 5.1 shows that on the first try the second girl could pick either the yellow (Y) or the white (W) ball. If she picked the yellow ball on the first try, she could pick either the yellow or the white ball on the second try. If she picked the white ball on the first try, she could also pick either the yellow or the white ball on the second try. So the sequence of possible events is:

1. Y and Y
2. Y and W
3. W and Y
4. W and W

There are four possible sequences. Since the sequence of events we want (Y-Y) is 1 out of 4 possible sequences, the chances of obtaining this sequence is 1 out of 4, or 1/4. Again, if we extrapolate to 100 cases, we would predict she would be correct about 25 out of 100 or 25% of the time. We would say her chance of being correct twice in a row is about 25%.

Questions

1. What is the chance (probability) of tossing a coin twice and having it come up heads both times?

   ____

2. What is the probability of rolling a die twice and having it come up with a six both times?

   ____

3. Suppose four balls are put into a sack. One of the balls is red and the other three are blue. What is the probability of reaching

in and picking a blue ball on the first try?

____

If you put the ball back in the sack after each pick, what are your chances of picking a blue ball again?

____

4. Last year the older sister of a biology student got married. She proclaimed that she and her husband were planning on having children right away. They both agreed that they wanted one girl and one boy. Assuming they will have two children, what are their chances of having one girl and one boy?

____

5. Local telephone numbers contain seven digits. Supposed you wanted to call a friend but you forgot the last two digits of his phone number. What is your probability of correctly guessing the last two digits on the first try?

____

The Mealworms in the Box

When asked to investigate the behavior of mealworms, Fritz decided to find out if mealworms respond to light. He took four worms and put them into the center of a box that had a neon light shining at one end. The other end was covered with black paper to shade it. Fritz figured that if the mealworms "liked" the light, the four worms would move to the lighted end. If they "liked" the dark they would crawl to the dark end. Fifteen minutes after putting the mealworms in the center of the box, he checked to see where the worms had crawled. The diagram in Figure 5.2 shows the results.

As you can see, one of the worms is in the light area and three are in the dark area. Does this mean that the worms prefer the dark? If so,

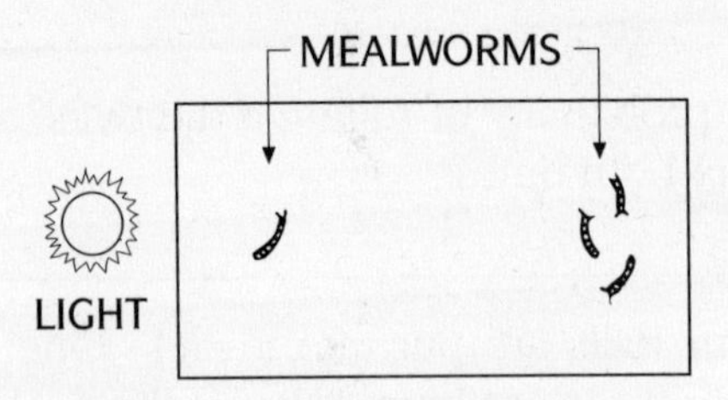

Figure 5.2 Mealworm response to light and dark.

why is one worm still in the light? Maybe it is just a strange worm. Or maybe it is not strange. Maybe Fritz's results are just due to "chance." Maybe if he waits a little while, all the worms will crawl to the lighted end. In other words, what is the probability that this result is due to chance alone and not due to the light?

How does one interpret such results? This is the central question in analyzing scientific data. To understand the answer, we need to understand more about probability.

Let's do the following activity as a way of better understanding the situation.

Draw pictures of the four mealworms on four pieces of paper. Draw each mealworm on both sides of the paper. Mark one side *L* for light and mark the other side *D* for dark. Put the four mealworms back into a sack and shake them up.

Turn the sack over and dump the worms on the table. How many worms have L's showing? __________ How many have D's showing? __________

Let's imagine that those with the L's showing are the ones that crawled to the lighted side of Fritz's box. The ones with D's showing are the ones that crawled to the dark side of the box. Record these numbers on Table 5.1 on page 110 next to Trial 1.

Put the four worms back into the sack and shake them up again. Dump the worms out on the table again and record on the data sheet the number of L's and D's that turned up this time next to Trial 2.

Repeat this procedure 30 more times.

How many times did four L's and zero D's turn up?

____

How many times did three L's and one D turn up?

____

How many times did two L's and two D's turn up?

____

How many times did one L and three D's turn up?

____

How many times did zero L's and four D's turn up?

____

TABLE 5.1 MEALWORM BEHAVIOR

	NUMBER OF WORMS IN			NUMBER OF WORMS IN	
TRIAL NO.	LIGHT SIDE	DARK SIDE	TRIAL NO.	LIGHT SIDE	DARK SIDE
1.			17.		
2.			18.		
3.			19.		
4.			20.		
5.			21.		
6.			22		
7.			23.		
8.			24.		
9.			25.		
10.			26.		
11.			27.		
12.			28.		
13.			29.		
14.			30.		
15.			31.		
16.			32.		

If these numbers were plotted on a graph that compares the five possibilities with their frequencies, what shape would the curve be? Why do you think the curve has that shape?

The numbers you obtained represent the outcome of an actual experiment. Let's compare those numbers with what would be expected to occur according to a tree diagram. The tree diagram would look like the one shown in Figure 5.3.

The diagram indicates that the first worm could have gone to the light (L) or the dark (D). Suppose it went to the light. The second worm could then have gone to the light or the dark. Suppose the second worm went to the light. The third worm could then have gone to the light or the dark and so on.

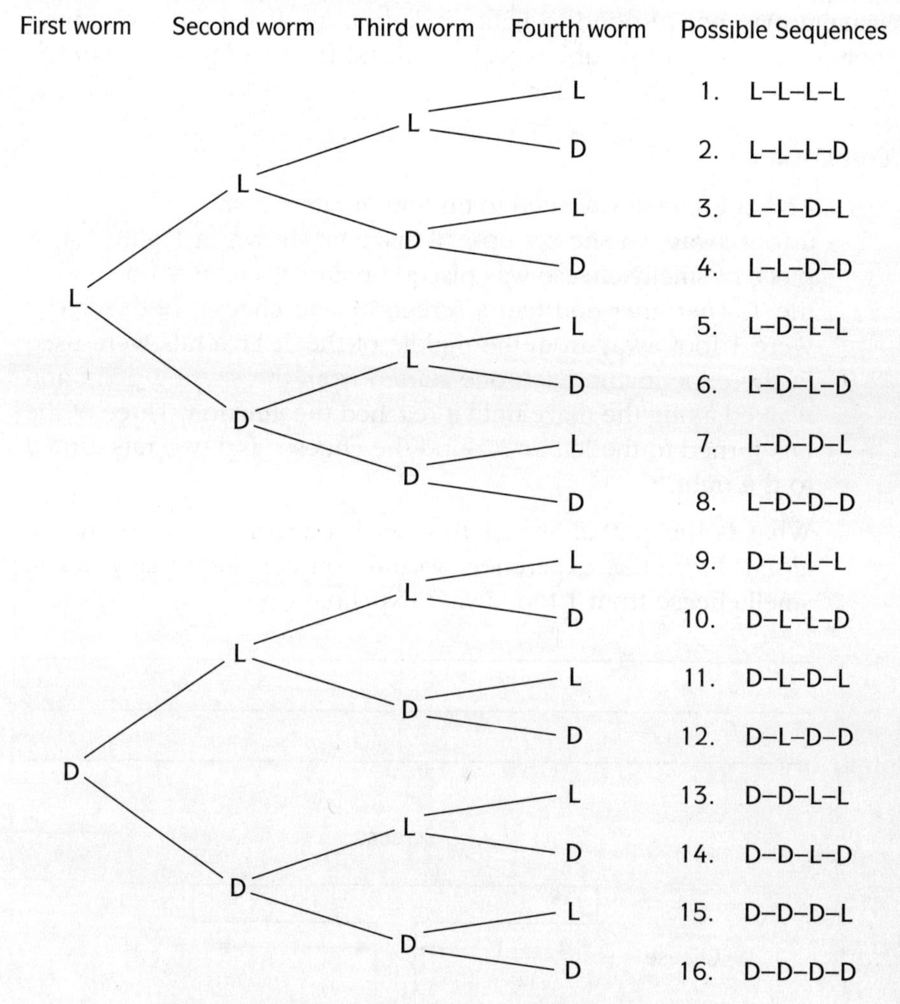

Figure 5.3 A tree diagram for mealworm responses.

As you can see, there are 16 possible sequences or ways in which the worms could have arranged themselves. Of these 16, only 4 of them have 1 worm in the light and 3 in the dark (numbers 8, 12, 14, and 15). So the probability of Fritz's result occurring just by chance is 4 out of 16 or 1 out of 4. In percentages this would be 25 out of 100 or 25%.

This is not a very large chance, so we could conclude that the dark (or absence of the light) was probably the cause of them being in that side of the box.

Notice, however, that we cannot be certain about this. If we conclude that mealworms prefer the dark, we could be wrong. The worms actually may not respond to light or dark at all. If we did the experiment four times, we would expect this result in one out of the four experiments just by chance alone.

Questions

1. An experimenter wanted to find out if rats can smell cheese from a foot away, so she set up a T maze as shown in Figure 5.4. A piece of smelly cheese was placed behind a screen at one end of the T. The other end had a screen but no cheese. Both screens were 1 foot away from the middle of the T. Five rats were used in the experiment. Each one started from the starting point and walked along the maze until it reached the junction. Three of the rats turned to the left and found the cheese, and two rats turned to the right.

   What is the probability of this result occurring due to chance alone? From this experiment would you conclude that rats can smell cheese from 1 foot away? Explain.

   ____

   ____

   ____

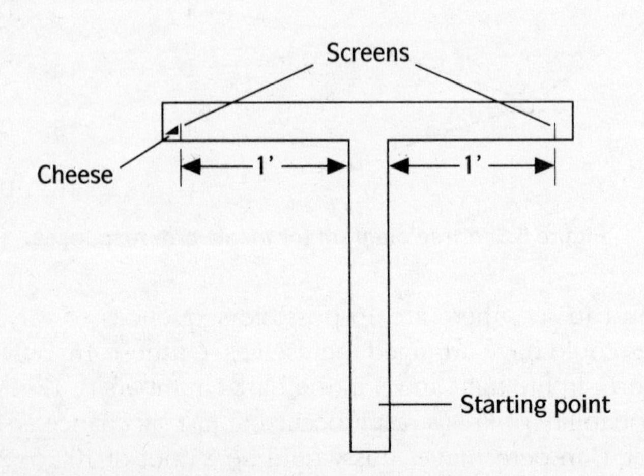

Figure 5.4 Experimental setup to test a rat's ability to smell cheese.

2. On a recent stop at the Valley Shopping Center, a man was giving a group of shoppers the "Cola Challenge." The Cola Challenge amounts to giving people a drink of Cola A and a drink of

Cola B without their knowing which is which. The people then report which tastes better to them.

As I watched, I noticed that three people said the glass that had Cola A in it tasted better. One person said she liked the glass that had Cola B in it better and two people said they could not tell the difference.

a. What fraction of the people preferred Cola A?

____

b. Suppose the Cola Challenge were given to more than one person. What would you say the chances are that she or he would prefer Cola A?

____

c. Does this sample of data show that people in general prefer Cola A to Cola B, or is the result just due to chance? Explain.

____

3. Last winter my boss held a dinner party at her house for the office group and their families. In all, 20 people were invited. Naturally, all 20 went. The boss served apple, cherry, and banana cream pie for dessert. On the day following the party, four of the five people who ate banana cream pie came down with severe stomach pains and nausea.

a. Does it seem likely that the banana cream pie was the cause of the stomach pains and nausea? Explain.

____

b. Can you figure out the probability of this many people getting sick to the stomach due to chance alone? If not, what other information do you need?

____

4. How many different license plates can there be in the state you live in? Generate a formula for finding the number of possible combinations.

____

____

____

5. You are on a television game show and given a choice of three doors. Behind one door is a new car; behind the others are pigs. You pick Door 1, and the host, who knows what is behind each door, opens Door 3, which has a pig. She then asks if you want to pick Door 2. Should you switch? Why? Why not? (Hint: You may be thinking that because two doors remain unopened with a car behind one and a pig behind the other, your chances are 1/2 for each door. Although to most people this seems reasonable, it is wrong! Generate all possible combinations of results before you jump to a hasty and incorrect conclusion.)

____

____

____

____

____

6. Suppose four isopods are placed into the center of a box with a light at one end. Ten minutes later all four of the isopods are found at the end away from the light. What is the probability that this result is due to chance alone?

____

   a. 1 out of 4

   b. 1 out of 8

   c. 1 out of 16

   d. 1 out of 32

   Draw a tree diagram in the space provided on the following page to determine the probability.

   Notice that only one out of the 16 possible outcomes (DDDD) shows all four isopods in the dark end (D). Therefore, the probability that this result is due to chance alone is one out of 16.

CONTROLLING VARIABLES

One day at the lake, two boys were overheard arguing about who could do more push-ups. The bigger boy said, "There is no way that you can do more push-ups than me. I'm bigger and stronger than you." To this the smaller boy replied, "Okay, let's see you prove it. You go first. I'll bet you a dollar that I can do more than you."

At hearing this, the bigger boy must have thought to himself, "What an easy way to win a dollar." He immediately got down on the beach and did 25 push-ups before he tired and could do no more. The smaller boy watched patiently and appeared not the least big shaken by the bigger boy's prowess.

At the conclusion of the bigger boy's effort, the smaller boy smiled and jumped into the water up to his calves. He got down in the water and proceeded to do 26 push-ups without breathing deeply. When he finished the bigger boy protested, "Hey, you can't do that. That's not fair. It's a lot easier to do push-ups in the water than on the beach!"

What do you think? Was it fair? Obviously not. One boy doing push-ups on the beach and the other in the water is not a fair test to find out who can do more push-ups. What would make it fair?

The attitudes and actions you develop as you conduct controlled experiments will allow you to judge and direct experience critically, rather than be led by it. The controlled experiment (fair test) will serve you, not only in the sciences, but in all efforts where you must evaluate alternative hypotheses and evidence.

The Tennis Balls Again

As the girl with the white tennis ball continued down the bumpy dirt road, she noticed a boy walking toward her. The boy was bouncing an orange tennis ball. At seeing the bouncing orange ball, she noted that it seemed to be bouncing a lot higher than her white ball.

After she said hello, the boy replied, "Hey, your tennis ball is a real dud. My orange ball is a lot bouncier than yours." Not wanting to be outdone, the girl responded, "Oh no it's not. Mine is bouncier than yours and I can prove it." To this the boy said, "Oh yeah, let's see you prove it. Go ahead. Drop both balls and let's see which bounces higher."

At this, the girl held her white ball over her head and the boy's orange ball near her waist and dropped them both at the same time. The white ball bounced higher!

To this the boy protested, "Hey, that's not fair. You can't drop them from different heights." So the girl took the two balls again and dropped them from the same height. But this time she dropped them so that the white ball hit a hard spot in the road and the orange ball hit a soft spot. The white ball bounced higher. Again the boy protested, "That's still not a fair test. My ball hit a soft spot and yours hit a hard spot. Do it again, but this time don't drop them from different heights and don't let one hit a soft spot."

The next time the girl did what the boy told her, but she released the balls so that the white one hit the sidewalk and the orange one hit the road. Again the white ball bounced higher than the orange ball.

By this time the boy was getting rather upset. He protested, "You're messing up the test. Don't drop them both from different heights. Don't let one hit a hard spot and the other a soft spot and don't let one hit the sidewalk." "Okay, Okay," exclaimed the girl. "Let me try again."

This time she held up the two balls and released them from the same height. They both hit hard spots in the road. But again the white ball bounced higher! This time she had cleverly spun the orange ball as she dropped it, so that when it hit the road it bounced at an odd angle and did not rise very high into the air.

At seeing this, the boy was so upset at the girl's failure to conduct a fair test that he grabbed his orange ball, turned around, and went off down the road muttering to himself.

Suppose you were the boy. What could you have said to the girl to keep her from messing up the test? You could of course do what the boy did and that is to tell the girl a number of things that she should not do (e.g., do not release the balls from different heights, do not let one hit a soft spot while the other hits a hard spot, do not let one hit the sidewalk while the other hits the road, do not spin one ball and not the other). If you did this, however, your sentence would get very long and she still might fool you. A better way is to simply tell the girl what she should do. And that simply is: Do the same thing to both balls!

If she had done the same thing to both balls (for example, released them from the same height, let them hit the same road surface, or dropped them with no spin), she would have conducted a "fair test." Tests have to be fair or else you cannot be sure if the results are due to the variable under consideration (in this case the difference in the two balls) or due to some misleading variable (the height of release, the surface the balls hit, the spin, etc.). Fair tests are so important in science that they have been given a special name. They are called controlled experiments. "Controlled" refers to the fact that the values of all the variables that might make a difference (except of course the variable under consideration) are kept the same. They are "controlled."

If you do a controlled experiment in which the values of only one variable are different (in this instance the balls themselves) and you get a difference in the result (in this instance one ball bounces higher than the other), then you can be sure that the difference is due to that variable (the orange ball really is bouncier). It must be due to that variable because all other variables were the same (both dropped from same height, hit the same surface, etc.).

The Pendulum

Let's see if you can conduct some fair tests (controlled experiments) to identify variables that make a difference in the number of times pendulums swing before stopping.

Start by obtaining the following materials:

thread

several paper clips

a table

tape

a watch with second hand or a stop watch

To construct a pendulum, cut a piece of thread about 18 inches long and tie one or more paper clips to one end. Tape the other end to the top of the table so that the thread and paper clips hang over the edge and can swing freely. Pull the paper clips to the side and let go so that the pendulum (the swinging thread and paper clips) swings back and forth. Time how long it takes for the pendulum to stop swinging.

You can also construct other pendulums and swing them in different ways.

Question: What variables do you think might affect the length of time it takes for different pendulums to stop swinging?

List of variables:

____

____

____

____

____

Perform controlled experiments to test at least three of these variables to see if they really do make a difference. For each experiment, write down the variable you are testing. Make sure that the values for the variables you are not testing do not change during the experiment. Why is this important?

___

What variables actually make a difference in the time it takes for a pendulum to stop swinging?

___

___

What variables do not make a difference?

___

___

Have you conducted controlled experiments? Explain.

___

___

___

___

___

In your experiments, the length of time the pendulum swings is the dependent variable because, as you found out, it is dependent upon another variable, the length of the thread. The length of the thread is an independent variable. Changing the length of the thread causes a change in the dependent variable. The longer the thread, the longer the time it takes the pendulum to stop swinging. Other independent variables that you might have tested are (1) the number of paper clips hung on the end, (2) the distance the pendulum was pulled back before it was released and (3) whether or not you gave it a shove when you released it. If you performed carefully controlled experiments of these variables, you should have discovered that the only independent variable that influences the dependent variable is thread length. Interestingly, many people think that the number of paper clips (the amount of weight at the end of the thread) also makes a difference, but it does not. Notice how doing controlled experiments can help people get rid of wrong ideas! This is very important because many people hold many "wrong" ideas. If people were better scientific thinkers, fewer wrong ideas would persist.

Questions

1. What was the dependent variable in the tennis ball experiments on pages 116–117?

____

2. What were the independent variables in the tennis ball experiments?

____

3. What was the dependent variable in the story about the push-ups?

____

4. What important independent variable was not controlled (i.e., kept the same) by the smaller boy in the push-up story?

____

5. Fifty pieces of various parts of plants were placed in each of 5 sealed jars of equal size under different conditions of color of light and temperature. At the start of the experiment, each jar contained 250 units of carbon dioxide. The amount of carbon dioxide in each jar at the end of the experiment is shown in Table 5.2. Which 2 jars would you select to make a fair compar-

TABLE 5.2 EXPERIMENTAL CONDITIONS AND RESULTS

JAR	PLANT TYPE	PLANT PART	COLOR OF LIGHT	TEMP. (°C)	CO_2*
1	Willow	Leaf	Blue	10	200
2	Maple	Leaf	Purple	23	50
3	Willow	Root	Red	18	300
4	Maple	Stem	Red	23	400
5	Willow	Leaf	Blue	23	150

*This column indicates cm^3 of CO_2 in the jars at the end of the experiment.

ison to find out if temperature makes a difference in the amount of carbon dioxide used?

______

6. In the previous chapter, the swimming coach did an experiment to find out who among those girls who tried out for the team were the fastest swimmers.

   a. What was the experiment?

   ______

   b. Was it a controlled experiment? Explain.

   ______

   c. What was the dependent variable (effect)?

   ______

   d. What independent variables (causes) are important in this situation?

   ______

7. Question 1 of the study problems in the previous chapter showed a graph of speed versus miles driven by automobile.

   a. What was the dependent variable (effect) in that situation?

   ______

   b. What was the primary independent variable (cause) investigated?

   ______

   c. What other independent variables might be important?

   ______

   d. What important independent variable was controlled?

   ______

   e. What important independent variable was not controlled?

   ______

f. How would you improve the experiment to find out if a correlation really does exist between speed and miles obtained?

___

8. A student recently conducted an experiment to find out if the color of the cup a person drinks from affects his or her perception of what he or she is drinking. The student predicted that people would prefer drinks from orange cups since his psychology teacher had remarked that orange was a warm color that had a soothing effect on one's disposition.

   He tested this prediction by putting identical kinds of and amounts of cola in two cups—one orange and one white. Wanting to do a controlled experiment, he proceeded in exactly the same way for each of the 15 persons he tested. First he told each person that they would be tested to find out which of two unknown kinds of cola they preferred. Each person was given an ounce of the first unknown cola in a white cup. Then each person was given a saltine cracker. Then each person was given an ounce of the second unknown cola (actually the same kind of cola) in an orange cup. He found that 11 out of the 15 people said they preferred the cola in the white cup!

   a. Was this a controlled experiment? Explain.

   ___

   b. Why do you think so many people chose the cola in the white cup?

   ___

   c. Does this experiment prove that the color of the cup affects a person's perception of what they are drinking? Explain.

   ___

   ___

   d. Are there any changes you would suggest to improve the experiment? If so, what are they?

   ___

   ___

   ___

   ___

PROPORTIONAL RELATIONSHIPS

If you have ever watched a good golfer, you probably noticed that the harder he or she swings the club, the farther the ball goes. Things are quite the opposite for most novices, however. The harder they swing, the shorter distance the ball goes. In fact, if the novice swings too hard, he or she may even miss the ball completely! Many other relationships like these exist in the environment. The farther away an object is, the smaller it appears. The more food people eat, the more weight they gain. The faster a car travels, the more gas per mile it uses. The longer a person's legs, the fewer steps he or she must take to get from the TV room to the refrigerator. The more a person exercises, the less his or her chances are of dying from heart disease.

As you know, such relationships are called correlations. The identification of correlations is crucial to understanding our environment. Not only is it important to be able to identify correlations, it is often important to be able to quantify such relationships. For example, everyone knows that a correlation exists between the amount of something you buy and the price you pay for it. The more you buy the more you pay. But do you know if you get a better deal when you buy 2 gallons of ice cream for $4.95 or 5 gallons of the same kind of ice cream for $6.75?

How High Will They Bounce?

Suppose the girl with the white tennis ball had actually conducted a fair test to find out whether her friend's orange tennis ball was bouncier than her white one. This would have been simple to do. She could have just dropped both balls under identical conditions and carefully noted which bounced higher. The one that bounced higher would clearly have been the bouncier ball.

If she had done the test, she would have found out that when the balls are dropped from waist high, the white ball bounces 1 foot into the air and the orange ball bounces 2 feet into the air.

Now, suppose the balls are dropped from near her shoulders and the white ball bounces 2 feet into the air. How high will the orange ball bounce under identical circumstances?

If you predicted 4 feet, you would have been correct. It seems that for every 1 foot the white ball rises, the orange ball will rise 2 feet. Therefore, when the white ball rises 2 feet, the orange ball will rise $2 \times 2 = 4$ feet.

Let's consider another example of this sort of quantitative relationship before we introduce a general method of dealing with such relationships.

Building Walls

Figure 5.5 contains blocks of various lengths and colors. Draw several of the red and white blocks on a piece of paper. Make sure you draw them the correct sizes. Let's see if you can construct some walls using the blocks.

Place one red block in front of you. Now place two white (noncolored) blocks side by side on top of the red block. Directly to the right of these blocks line up four red blocks end to end as shown.

Suppose you were to place white blocks on top of these red blocks to complete the wall. How many white blocks would you need?

The answer of course is eight. Since there are two white blocks for every one red block, there would be eight white blocks for the four red ones. Construct the wall if you would like to verify this answer. As you probably know, the relationship between the two walls can be written symbolically as follows:

$$\frac{\text{2 whites}}{\text{1 red}} = \frac{\text{8 whites}}{\text{4 reds}}$$

or

$$\frac{2}{1} = \frac{8}{4}$$

What we have are two ratios that are equal. This sort of a mathematical relationship is known as a *proportion*.

Now cut out and place two dark green (dg) blocks end to end on the table in front of you. Place three purple (p) blocks end to end on

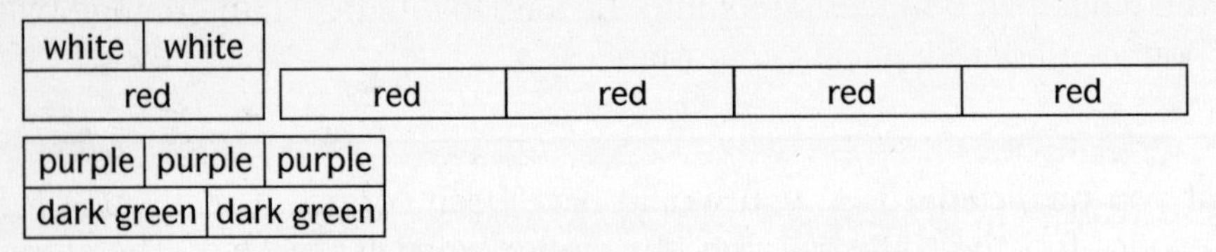

Figure 5.5 Visualizing quantitative relationships.

top of the dark green blocks. As you see, there are three purple blocks for every two dark green blocks (a ratio of 3(p)/2(dg)).

Directly to the right of these blocks, place six dark green blocks end to end. How many purple blocks would you need to place on top of these dark green blocks to complete the wall?

If you said nine, you are correct. Symbolically, the problem looks like this:

$$\frac{3(p)}{2(dg)} = \frac{?(p)}{6(dg)}$$

One way to solve the problem is simply to note than 2(dg) × 3(dg) = 6(dg) and that 3(p) × 3(p) = 9(p). Or you could divide 2(dg) into the 6(p) to give 3. Then multiply this 3 by the 3(p) to give 9(p). Construct the wall if you would like to verify this answer.

Let's start again with three purple blocks stacked on top of two dark green blocks. Directly to the right of this wall, place nine dark green blocks end to end. How many purple blocks would you need to place on top of these nine dark green blocks to complete the wall?

This time the wall cannot be built with a perfect matching of the blocks. You would need 13.5 purple blocks to match the nine dark green ones.

Using the same symbols, this problem would look like this:

$$\frac{3(p)}{2(dg)} = \frac{?(p)}{9(dg)}$$

One way to solve this problem is to divide the two dark green blocks into the nine dark green blocks, giving an answer of 4.5 times. Then simply multiply this 4.5 by the three purple blocks, which results in 13.5 purple blocks.

$$9(dg) \div 2(dg) = 4\ 1/2,\ 4\ 1/2 \times 3(p) = 13\ 1/2(p)$$

Again construct the wall if you would like to verify this answer.

If you are somewhat unsure of how to solve problems like these, make up a few problems for yourself. Try using just a pencil and paper to solve the problems and then verify your solutions by actually building the walls.

The Coupled Gears

To provide a different experience with proportional relationships, you might like to try to solve some problems with a set of coupled gears. Start by tracing, cutting out, and assembling the gears as shown in Figure 5.6 . Use pins to hold the gears in place on a piece of cardboard.

Experiment 1 Rotate the smaller gear three times by turning the shaft. How many times did the larger gear turn? Use Table 5.3 to record your data.

Experiment 2 Suppose the smaller gear is rotated six times. How many times do you predict the larger gear will rotate?

Check your prediction by performing the experiment. Record the result in Table 5.3.

Figure 5.6 Using gears to visualize proportional relationships.

TABLE 5.3 RECORDING DATA ON GEAR TURNING

	EXPERIMENT NO.					
NO. OF TURNS ON GEAR	**1**	**2**	**3**	**4**	**5**	**6**
Smaller gear	3	6		78		2
Larger gear			6		5	

Experiment 3 How many times do you predict the smaller gear will rotate when the larger gear is rotated six times?

Check your prediction by performing the experiment. Record the result.

As you can see, a direct correlation exists between the number of rotations of the two gears. The more one gear rotates, the more the other gear rotates. In fact, the small gear rotates three times for every two times the large gear rotates. Since the numerical relationship that exists consists of equivalent ratios—for example:

$$\frac{3(s)}{2(l)} = \frac{6(s)}{4(l)} = \frac{9(s)}{6(l)}$$

the relationship is said to be directly proportional.

Experiment 4 Suppose the smaller gear is rotated 78 times. How many times do you predict the larger gear will rotate?

Actually performing this experiment will be difficult if you are working by yourself. It would be tedious and you might lose count. Do you actually need to do the experiment to be sure of your answer?

The problem can be set up like this:

$$\frac{3(s)}{2(l)} = \frac{78(s)}{?(l)}$$

. . . and solved like this:

$$\frac{3(s) \times 26}{2(l) \times 26} = \frac{78(s)}{52(l)}$$

The 26 was obtained by dividing 78 by 3.

Experiment 5 How many times do you predict the small gear will rotate when the large gear is rotated five times?

____

The problem can be set up like this:

$$\frac{3(s)}{2(l)} = \frac{?(s)}{5(l)}$$

Check your prediction by performing the experiment. Record the result.

____

____

____

____

Experiment 6 Suppose the smaller gear is rotated only two times. How many times do you predict the larger gear will rotate?

____

Show how you set up the problem and arrived at your prediction.

____

____

____

____

Check your prediction by performing the experiment. Record your result.

____

____

____

____

____

____

Balancing Weights

Thus far the correlations we have investigated have all been direct; that is, as the value of one variable increased, so did the value of the second variable. But as you know, some correlations are inverse in the sense that as one variable increases the other variable decreases. Let's see how the variables of distance and weight are correlated in a balancing beam.

Fasten a piece of string to the exact center of a yardstick and fasten the other end of the string to the edge of a table so that the yardstick hangs freely and is balanced horizontally below the table. Now tape a metal washer on one side of the yardstick four inches from the center. Where would you tape another washer of equal weight to make the yard stick balance?

____

If your first guess does not make the yardstick balance, keep trying until you are successful.

Now try this series of experiments. In each experiment, test your prediction by hanging the second set of washers on the beam.

1. Hang two washers four inches to the right of center. Predict where you would hang one washer to the left to make the beam balance.

   ____

2. Hang one washer six inches to the left. Predict where you would hang two washers to the right to make the beam balance.

   ____

3. Hang one washer three inches to the right. Predict where you would hang three washers on the left.

   ____

4. Hang two washers three inches to the left. Predict where you would hang three washers on the right.

   ____

5. Hang five washers six inches to the right. Predict where you would hang three washers on the left.

____

As you have no doubt noticed, the relationship between the weight and distance is not a direct one. In fact, the less weight you have the further it has to go out toward the end to make the beam balance. An inverse correlation exists. But how can problems like these be solved?

Let's look at the first problem. You were given one-half of the original weight. To balance the beam, you had to hang it twice as far out as the original weight was hung. The original weight was hung at four inches from the center, so the lighter weight had to be hung at eight inches from the center ($2 \times 4 = 8$). In short, it is one-half the weight, so it goes twice as far out. Symbolically, the problem looks like this:

1. $$\frac{2\text{(weight on right)}}{1\text{(weight on left)}} \quad \frac{8\text{(distance on left)}}{4\text{(distance on right)}}$$

   The crossed arrows indicate the inverse relationship. The weights and distances are said to be related in an inverse or *reciprocal* fashion.

   Using the same procedure, the remaining problems can be set up as follows:

2. $$\frac{1\text{(weight on left)}}{2\text{(weight on right)}} \quad \frac{?\text{(distance on right)}}{6\text{(distance on left)}}$$

   It is twice (2/1) the weight, so it must be hung at one-half (1/2) the distance,

   $1/2$ of $6 = 1/2 \times 6/1 = 6/2 = 3$.

3. $$\frac{1\text{(weight on right)}}{3\text{(weight on left)}} \quad \frac{?\text{(distance on left)}}{3\text{(distance on right)}}$$

   It is three times (3/1) the weight, so it must be hung one-third (1/3) the distance,

   $1/3$ of $3 = 1/3 \times 3/1 = 3/3 = 1$.

4. $$\frac{2\text{(weight on left)}}{3\text{(weight on right)}} \quad \frac{?\text{(distance on right)}}{3\text{(distance on left)}}$$

It is three halves (3/2) the weight, so it must be hung two-thirds (2/3) the distance,

$2/3 \text{ of } 3 = 2/3 \times 3/1 = 6/3 = 2.$

5. 5(weight on right) / 3(weight on left)  ?(distance on left) / 6(distance on right)

It's three-fifths (3/5) the weight, so it must be hung five thirds (5/3) the distance,

$5/3 \text{ of } 6 = 5/3 \times 6/1 = 30/3 = 10.$

Here are a few additional problems you might like to try to be sure that you understand the relationships.

6. Hang three weights at number ten on the right. Where would you have five weights on the left?

____

7. Hang four weights at number two on the left. Where would you hang two weights on the right?

____

8. Hang three weights at number eight on the right. Where would you hang four weights on the left?

____

9. Hang seven weights at number five on the left. Where would you hang four weights on the right?

____

10. Hang five weights at number seven on the right. Where would you hang seven weights on the left?

____

QUESTIONS

1. The graph in Figure 5.7 on page 132 shows that as the dose of x-radiation increases, so does the number of mutations in organisms. The relationship is directly proportional. For every five relative units increase in dosage of x-radiation, four more mutations occur.

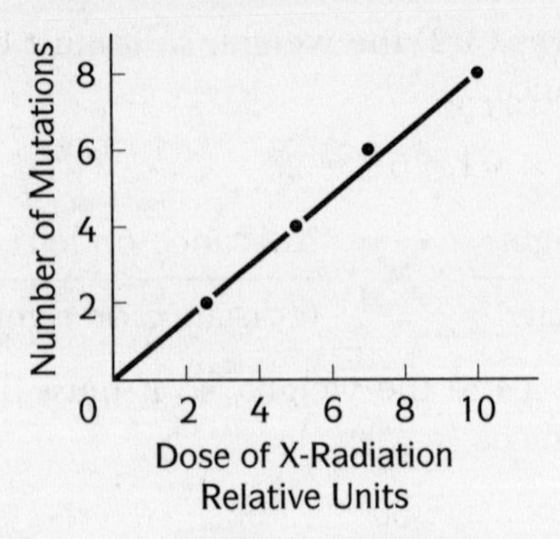

Figure 5.7 The proportional relationship between a dose of x-radiation and the number of mutations in organisms.

a. Suppose the organisms were given a dose of 30 relative units of x-radiation. How many mutations would you predict would occur?

____

Hint: $\frac{5\,\text{ru}}{4\,\text{m}} = \frac{30\,\text{ru}}{?\,\text{m}}$

b. How many mutations would occur if the dosage were 8 relative units?

____

2. The velocity at which a nerve impulse moves is directly proportional to the diameter of the nerve fiber. The velocity is about 10 miles per hour for every 2 microns diameter of fiber. At about how many miles per hour will a nerve impulse travel down a nerve 16 microns in diameter? 9 microns in diameter?

____

3. A triple recipe for cookies calls for 4 cups of milk and 5 eggs. How many cups of milk and eggs should you use for a double recipe?

____

4. John is 6 years old and his sister Linda is 8 years old. When John is twice as old as he is now, how old will Linda be?

____

5. The time it takes for two types of molecules to move across a cell membrane is directly proportional to the square roots of their molecular weights. In an experiment that compares movement of molecule A (molecular weight 720) and molecule B (molecular weight 370), molecule A is detected inside the cell about 6 seconds after its release into the extracellular fluid. How long after release would you expect molecule B to be detected in the cell?

____

6. A department store is having a sale on socks, six pairs for $4.00. Another department store is also having a sale on the same kind of socks, eight pairs for $5.25. Suppose both department stores were an equal distance from your house. At which store do you think you would get a better deal on the socks?

____

7. A woman must mix oil and gasoline for a lawn mower. She must add 1/2 pint of oil for each gallon of gas.

   What are the two variables in this situation?

   ____

   What is the relationship between these variables?

   ____

   Does the difference between them stay the same (i.e., constant difference) or is the relationship proportional?

   ____

8. Sam and Cindy are parachuting toward the earth at the same velocity. Sam jumped out of the plane after Cindy, so he is above her. What are the two variables in this situation?

   ____

   What is the relationship between them (i.e., constant difference, proportion)?

   ____

9. A student is measuring the heights of various objects and the lengths of their shadows at a certain time of day. What are the two variables?

___

What is the relationship between them?

___

10. A distance of 16 kilometers in the metric system is equivalent to a distance of 10 miles. How many kilometers are equal to 35 miles?

___

11. In spring, the depth of water in the shallow part of a pond is 2 feet, while at the deep end it is 8 feet. During the summer, water has evaporated and the depth of water in the shallow end is only 6 inches. How deep is it in the summer in the deep end?

___

12. The unit of currency in Switzerland is the Swiss franc. Fourteen Swiss francs are worth 4 U.S. dollars. How many francs would you receive for 6 dollars?

___

13. Eratosthenes of Alexandria (273–192 BC), given the following information and assuming the Earth is round and the sun is so far away that its rays strike the Earth parallel to each other, accurately computed the distance around the Earth:

    - The distance from the city of Alexandria to the city of Syene is 800 km.
    - Alexandria is directly north of Syene.
    - At noon on June 21 a post casts no shadow at Syene. (Note: the post is pointing toward the center of the Earth.)
    - At noon on June 21, a similar post casts a shadow at Alexandria. (This post also points directly toward the center of the Earth.)
    - The angle at which lines drawn from the posts to the center of the Earth intersect = 7°.

    See if you can do what Eratosthenes did and figure out the distance around the Earth.

    ___

CHAPTER 6

THEORIES ABOUT THE NATURE OF LIFE

GETTING FOCUSED

- *Learn how hypotheses about the nature of life are tested.*
- *Learn about a modern theory of life.*
- *Learn about a theory of biological organization and emergent properties.*

The thinking patterns you read about in Chapters 2–5 were used in attempts to answer descriptive and causal questions about living things. Over the years, this resulted in the establishment of several generally accepted biological theories.

These theories represent modern "beliefs" among informed biologists and provide the "structure" of present day biological thought. These major theories are presented in the remaining chapters in a brief and straightforward manner so that you can quickly gain an understanding of the most important modern biological concepts and theories. Also included are several examples of the scientific thinking patterns that led to the development of the theories so that you can gain an even better understanding of their use.

This chapter begins with a discussion of some very famous experiments from the past that tested some key hypotheses about the nature of living things. This is followed by presentation of a theory of life and a theory of biological organization.

DO LIVING THINGS CONTAIN A "VITAL" FORCE?

Most people agree that things, living and nonliving alike, are made up of combinations of tiny nonliving particles called atoms, and that nonliving things are controlled by physical forces. But what about living things? Are they controlled by the same physical forces that control nonliving things? Or do they contain something extra—an unexplained, "vital" (life) force that somehow goes beyond mere physical forces? Such a belief, called **vitalism,** was expressed by a number of people in the past. How would one go about testing to see if such a vital force exists in living things?

In 1748 an English clergyman named John Needham reported the results of an experiment that seemed to show that life can arise from nonliving matter spontaneously (that is, without help from the experimenter). In that experiment, Needham put some mutton gravy into a bottle that he then plugged with a cork to make sure that no tiny living things or their eggs would enter from the air. Next, he heated the gravy in a hot fire. He believed this would kill any living things or eggs that might be in the bottle. He put the bottle away for a few days. When he returned to examine the gravy under a microscope, he found it teaming with tiny living creatures. Needham concluded from his experiment that living things can arise spontaneously from the nonliving. To explain how this could happen, he postulated that a special vital, mystic (mysterious) force entered the bottle and acted upon nonliving material to bring it to life. This theory was called **spontaneous generation.**

An Italian physiologist named Lazarro Spallanzani read about Needham's experiment and about his idea of the vital force. Spallanzani did not believe in spontaneous generation or in the vital force, so he looked for a flaw in Needham's experiment. To refute Needham's work, Spallanzani began with the alternative hypothesis that microbes grew in Needham's bottle because he had not sealed it tight enough to keep new microbes out and because he had not heated the gravy long enough to kill the microbes or eggs that were in the gravy at the start.

So Spallanzani took several bottles and seeds. He carefully cleaned the bottles and put some seeds in each. Following this, he poured distilled water into the bottles and then boiled the contents. To close the bottles, he melted the necks in a flame. He boiled some for only a few minutes and some for as long as an hour. As a control, he repeated the procedure with another set of bottles, except with these he plugged the tops with corks; he did not melt the necks.

Days later he returned to observe under his microscope what had happened. In the bottles that had been boiled only a short time, he found microbes; however, in those boiled for an hour and sealed by melting the necks, he found none. The bottles he had corked as Needham had, he found to be full of microbes, even the ones boiled for an hour.

The reasoning that guided Spallanzani's experiment can be summarized as follows:

Hypothesis: *If*...a vital force exists that can act on nonliving matter to bring it to life...

and...some bottles are heated and corked while others are heated and sealed by melting their necks...

Prediction: *then*...microbes should be spontaneously generated in both sets of bottles.

On the other hand,

Alternative Hypothesis: *If*...the vital force does not exist, and microbes can enter a bottle around a cork but not through a bottle neck that has been melted shut...

Prediction: *then*...microbes should be found in the corked bottles but not in the melted shut bottles.

Because Spallanzani's experimental results did not match the result predicted by the vital force hypothesis but did match the result predicted by his alternative hypothesis, he concluded that the vital force/spontaneous generation hypothesis was wrong. Instead, he believed that his alternative **biogenesis** hypothesis (life only from prior life) was correct.

To Spallanzani, his experiment seemed to be conclusive evidence that spontaneous generation and Needham's vital force did not exist. But now a curious thing happened. Instead of Needham admitting that he was wrong, he answered Spallanzani in a way that to many was quite convincing. Needham reasoned that the seeds contained the vital force and that the fierce heat used in Spallanzani's experiment weakened and so damaged the vital force that it could no longer create the microbes.

Spallanzani was, no doubt, very frustrated by Needham's latest claim, but again he saw the flaw in Needham's argument and attempted to point it out experimentally. Once again the Italian got out his equipment. He set up a whole series of bottles with seeds that had been heated for various lengths of time. The first seeds were boiled only a few minutes, some for half an hour, some for an hour, some for two hours, and some he baked until they were charred. After this he set up his bottles with the seeds and water, but he sealed the tops only with corks. If Needham was correct, no microbes would be found in the flasks with the charred seeds. Days later Spallanzani examined the flasks and found them all alive with microbes.

Surely Needham would not be able to argue against this evidence of this crucial experiment. But argue it he did, for he had made the ingenious claim that, while Spallanzani was heating his bottles, he was destroying the "elasticity" of the air. According to Needham, "elastic" air was necessary for the vital force to work.

Spallanzani doggedly set out to disprove this idea. All the bottles he had used previously had wide necks, so heating to seal them required a relatively long time. This, he reasoned, heated and as a result drove out a large quantity of air, which to Needham had appeared to be the result of boiling the sealed bottles, therefore his claim of less elastic air. Spallanzani instead took the same bottles and filled them partially with seeds and water. He then, by heating, reduced the necks of the vessels until the openings were very thin. After letting the internal and external air come to the same temperature, he put the openings to his blowpipe to seal them instantaneously so that the internal air was not changed.

With this completed, he heated the bottles in boiling water for an hour. Upon opening of the bottles nearly a month later, Spallanzani held a candle near the opening and found that the flame curved away from the neck. This showed that the internal air was more "elastic," not less, as Needham had argued.

Again a crucial experiment had been conducted. Needham and his supporters surely would have nothing with which to confront this. Spallanzani had shown tremendous skill in deducing the importance of his beliefs and putting them to the test. Indeed, much of the opposition to Spallanzani quieted, but it should be noted that the ideas of spontaneous generation and special vital forces in living things were by no means dead. Many distinguished scientists still held fast to earlier beliefs.

The question of spontaneous generation and vitalism persisted well into the nineteenth century, as witnessed by the classic experiments of Louis Pasteur, in which he concluded: "The great interest of this method is that it proves without doubt that the origin of life, in infusions which have been boiled, arises uniquely from the solid particles which are suspended in the air." The method Pasteur referred to was simply to set up bottles with nutrient liquid as others had done before. However, instead of sealing them, he drew the necks out under a flame, so that a number of curves were produced in them and some section of each neck was pointed downward (see Figure 6.1 on page 140). He then boiled the liquid for several minutes and allowed the flasks to cool and sit for several days. In no case did microbes develop in the liquid. This to Pasteur and many others seemed crucial, just as Spallanzani's experiments nearly a century earlier had. But even in the face of Pasteur's logic, clarity, and simplicity in experimentation, the question was still not completely settled in some people's minds.

A MODERN THEORY OF LIFE

Experiments like those of Spallazani, Pasteur, and others have led biologists to construct a modern theory of life. That theory argues that living things (organisms), like nonliving things, are composed of various combinations of atoms and molecules in motion. Physical forces act on these atoms and molecules in the same way in both living and nonliving things. Living things are composed mainly of carbon, hydrogen, nitrogen, oxygen, phosphorous, and sulfur atoms (CHNOPS). These

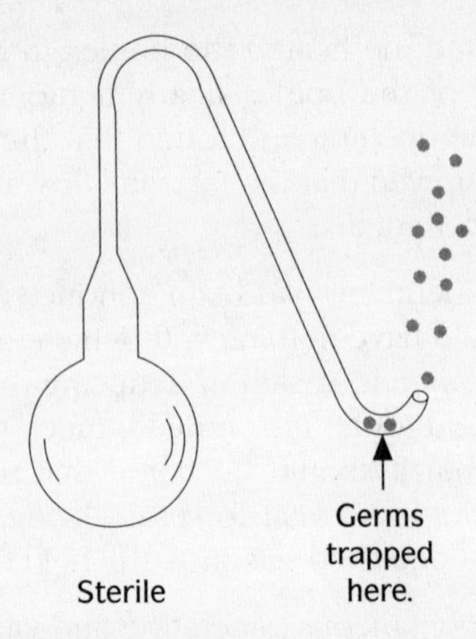

Figure 6.1 Diagram of Pasteur's flask with a section of the neck pointing down so that airborne microbes (germs) could not enter the flask even though the flask was open to the air.

atoms combine into a few kinds of complex molecules, such as proteins, carbohydrates, fats, and nucleic acids. Between 70%–90% of all living things is water.

Living things are distinguished from nonliving things mainly by their ability to grow, develop, reproduce, and die; they are chemically complex and they require energy to preserve their chemical complexity; their forms may change across generations; they contain a set of instructions that direct their activities; they are capable of responding to environmental stimuli; and they are composed of one or more normally tiny compartments called cells.

The line between living and nonliving things, however, is not always clear. Some objects, such as viruses, possess some but not all of these characteristics. **Viruses** consist primarily of chemically complex nucleic acid and protein molecules. The nucleic acid molecules serve as a set of instructions that direct the activity of the virus. When injected into a living cell, the molecules can direct the cell to produce many replicas (copies) of the virus. Viral forms are also capable of change across generations. However, if left in a dry bottle on a shelf, viruses will form crystals as salt or sugar molecules do, and they will not grow, develop, reproduce, or die. These characteristics suggest that viruses should be considered neither living nor nonliving. They are clearly intermediate (in between).

POSTULATES OF MODERN THEORY OF LIFE

1. Living and nonliving things alike are composed of atoms controlled by physical forces.
2. Living things (organisms), unlike nonliving things, possess the ability to grow, develop, reproduce, and die; they are chemically complex, and energy is needed to preserve this complexity; they contain a set of instructions that directs their activities; they are capable of responding to environmental stimuli; and they are composed of one or more cells.
3. Objects, such as viruses, exist that possess some, but not all, of the characteristics of living things, thus are intermediate.

A THEORY OF BIOLOGICAL ORGANIZATION AND EMERGENT PROPERTIES

Earth contains both living and nonliving things. Scientists have found it useful to think of the living and nonliving things on Earth as being organized into distinct levels of complexity. The simplest and least complex level consists of the **subatomic particles**—the protons, neutrons, and electrons that combine to make up **atoms.** Atoms are the smallest units of a substance that retain the properties of that substance. Ninety-two kinds of atoms have been found to occur naturally on Earth. Complexity increases as atoms of various kinds combine with one another to form units called **molecules** (i.e., two or more atoms bonded together). Under appropriate conditions, various kinds of molecules can in turn combine to form still more complex units called **organelles.** Figure 6.2 on page 143 shows the following additional levels as complexity continues to increase to the largest, most inclusive level of complexity on Earth, the biosphere:

cells A variety of organelles may interact with one another and form a still larger and more complex unit called a cell. Cells are the smallest units that are themselves normally considered to be alive.

tissues Group of similar cells that perform a specific function (e.g., muscle tissue and red blood cells in animals, bark tissue in plants).

organ Group of tissues that together perform still more complex functions, (e.g., brain, heart, and stomach in animals; roots, leaves, and stems in plants).

organ system Group of functionally related organs (e.g., digestive system in animals, consisting of such organs as the esophagus, stomach, and intestine; the root system in plants).

multicellular organisms Group of organ systems that function together to carry out all necessary life activities.

populations Group of similar kinds (species) of organisms that live and reproduce in a particular location.

biological community Populations of different species of organisms in a particular location that interact with one another.

ecosystem The populations within a biological community interact with one another, and they also interact with the nonliving components of their environment (e.g., sunlight, air, water, soil) to form complex units known as ecosystems (e.g., lake ecosystem, desert ecosystem, forest ecosystem).

biosphere All of the world's ecosystems interact in direct and/or indirect ways with one another in the all-inclusive and most complex unit of organization on Earth: the biosphere.

Thinking of the living and nonliving things on Earth in terms of these levels of organization is very useful. Typically, scientists investigate only a few levels at a time. A nuclear physicist, for example, tries to understand the behavior of protons, neutrons, and electrons, and what they are made of. A cell biologist tries to understand how the organelles within cells behave and what they consist of, while an ecologist may be trying to discover the factors that are causing an eagle population to decline. Investigations take place at different levels, with the ultimate objective of understanding each level and how the levels themselves interact. For example, to "understand" nature at the population level, one must understand how the "units" at the next lower level interact, that is, individuals, but one does not necessarily have to investigate interactions of units at still lower levels, for example, organs, cells, and atoms.

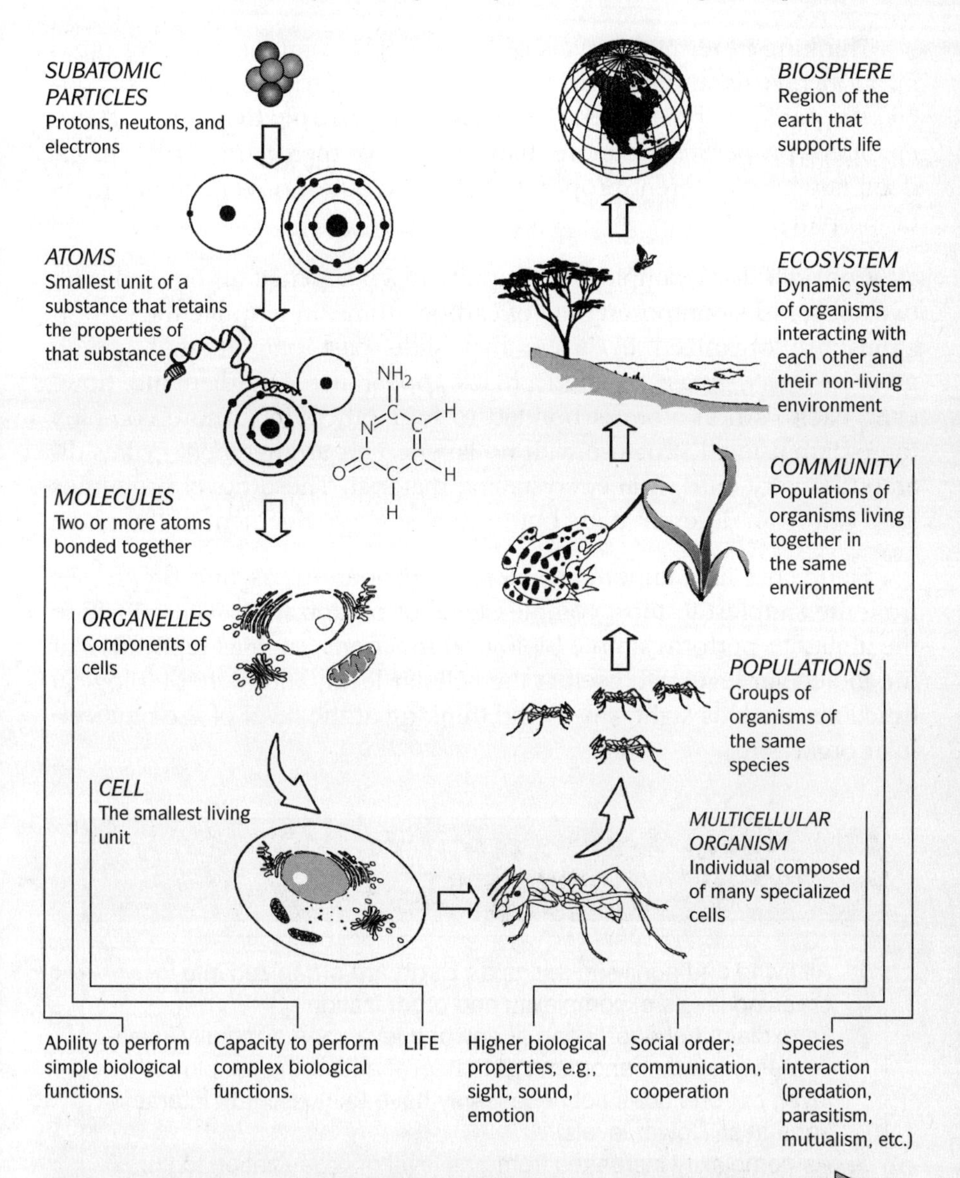

Figure 6.2 The major levels of organization, including some of the unique properties that emerge as the level of complexity increases.

Thinking in terms of levels of increasingly complex levels of organization reveals another extremely interesting fact. As complexity increases, unique phenomena or **emergent properties** arise. These emergent properties are more than the sum of their parts. They are indeed new and different properties that arise out of novel combinations of the parts.

Consider, for example, the graphite in your pencil and a diamond, two substances composed only of carbon atoms. In graphite the carbon atoms are organized in layers that slide past one another, giving graphite a soft, greasy feel and a black appearance. In a diamond, however, each carbon atom is bonded to four others in a more complex three-dimensional structure with no layers. This arrangement makes diamond a very hard, brittle, crystalline material. These novel properties arise not from different parts but from a new arrangement of the parts.

Figure 6.2 lists important emergent properties, as one progresses from the simplest to most complex level of organization. These include the ability to perform simple biological functions, such as replication at the molecular level; life itself at the cellular level; and higher biological functions, such as sight, smell, and thinking at the level of the multicellular organism.

POSTULATES OF A THEORY OF BIOLOGICAL ORGANIZATION

1. All living and nonliving things on Earth are organized into increasing levels of complexity and organization.
2. To explain a phenomenon at any one level, one generally investigates related phenomena at that level and at the next lower level, but one does not necessarily have to investigate interactions at still lower levels.
3. As complexity increases from one level of organization to the next, new properties appear that are more than the sum of their parts.
4. Emergent properties arise from new arrangements of their parts, not from new parts.

QUESTIONS

1. The words *prove* and *disprove* are sometimes used to refer to the results of scientific experiments; that is, the experiments prove that substance *x* causes cancer in rats. Look up the meaning of *prove* in a dictionary. Do you think scientific evidence can ever really prove or disprove any particular hypothesis? Explain.

____

____

____

____

____

2. Consult a mathematics textbook and/or a mathematics instructor to define *prove* as it is used in mathematics. How is this definition similar to or different from its scientific definition?

____

____

____

____

____

3. Compare a child at birth to that same child at age five years. Create a list of at least four behaviors (i.e., emergent properties) that have emerged during that child's first five years of life. Name two behaviors present at both ages.

____

____

____

____

____

4. Into what levels of organization should the phenomena that exist between the biosphere level and the most inclusive level, the universe, be classified? Explain.

5. Suppose you and a friend observe a painting of a large dog. You think the painting is beautiful but your friend thinks it is ugly. Develop a hypothesis for this difference of opinion. At what level of organization is your hypothesis? Does one need to understand how photons of light interact with one's eyes to understand your hypothesis? Explain.

6. Suppose you wish to understand why the population of bald eagles in Montana has been declining over the past three years. At what level of organization would you most likely discover an answer (e.g., the atomic level, the tissue level, the individual level, the biosphere level)? Explain.

TERMS TO KNOW

atom	organ
biogenesis	organelle
biological community	organ system
biosphere	population
cell	spontaneous generation
ecosystem	subatomic particles
emergent property	tissue
molecule	virus
multicellular organism	vitalism

SAMPLE EXAM

Based on Figure 6.3, circle the best answer for each question.

Flasks 1 through 8 were half filled with a liquid known to support the growth of microorganisms. Cotton stoppers were applied as indicated. Group I was treated no further and was allowed to stand at room temperature on a laboratory table. Group II was sterilized under pressure and placed next to Group I.

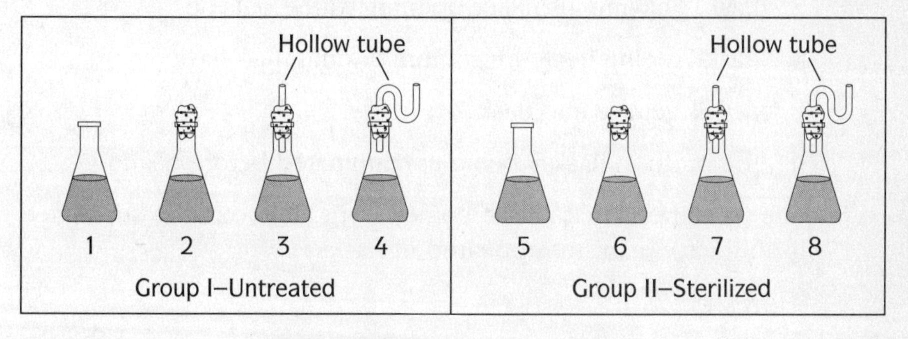

Figure 6.3 Experimental setup.

1. Flask 1 can serve as a control (i.e., a comparison) for
   a. flasks 2 and 6.
   b. flasks 5 and 6.
   c. flasks 7 and 8.
   d. flasks 2 and 5.
2. Organisms are most likely to appear first in flask
   a. 5.
   b. 6.
   c. 7.
   d. 8.
3. If all organisms are killed by sterilization and if airborne microbes and/or their spores cannot pass through cotton, no growth should appear in flask
   a. 4.
   b. 5.
   c. 6.
   d. 7.
4. If flask 5 became cloudy before flask 7, this cloudiness is most likely due to
   a. flask 7 having a smaller opening to the outside.
   b. flask 7 being heated for a longer time than flask 5.
   c. no air getting into flask 7.
   d. the broth in flask 5 being contaminated before heating.
5. The hypothesis that all life comes from life would be supported if no microorganisms appeared in flasks
   a. 3 and 5.
   b. 4 and 5.
   c. 5 and 7.
   d. 6 and 8.

6. Vitalism (i.e., *a*biogenesis) would be best supported if microorganisms grew in flasks

   a. 1 and 2.

   b. 3 and 4.

   c. 5 and 6.

   d. 6 and 8.

7. The hypothesis that microorganisms enter the broth only from the air would be best supported if microorganisms grow in flasks

   a. 1 and 8.

   b. 5 and 6.

   c. 5 and 7.

   d. 6 and 8.

8. If *no* microorganisms grew in flask 8, these data would support the hypothesis of

   a. Spallanzani.

   b. Pouchet.

   c. Needham.

   d. Aristotle.

CHAPTER 7

ORGANISM LEVEL THEORIES

GETTING FOCUSED

- *Understand the theories of organism classification and organism behavior.*
- *Learn how hypotheses about the origin and behavior of organisms are tested.*
- *Evaluate Mendel's theory of inheritance.*

Biologists are a diverse group of people who study vastly different aspects of nature. A cell biologist explores nature at the cellular level; a taxonomist explores nature at the organism level; while an ecologist might explore an entire ecosystem. Textbook authors and biology instructors who want to give students an overview of the entire field of biology are faced with the problem of how best to organize and present the vast subject matter. Should the micro-to-macro approach be used, starting at the atomic and molecular level going straight up to cells tissues and so forth, and ending up with ecosystems and the biosphere? Or should the approach start with the biosphere and go in reverse order? How about beginning in the middle, then go in either direction, and so on?

There is probably no one best approach, but in this textbook the remaining chapters begin at the organism level and progress down to the level of organ's systems, organs, tissues, cells, molecules, and atoms. The organism and population levels will be returned to in Chapter 12 and progress up to theories on the ecosystem level. Finally, the last chapter considers theories of the origin and evolution of life. This approach has been taken for three primary reasons. First, you are an organism and are most familiar with life at that level. Starting with the familiar and progressing to the unfamiliar should increase your understanding. Second, this approach allows some of the nature of inquiry to be used in these chapters. The inquiries of biologists in the past have also progressed from the familiar and known to the unfamiliar and unknown. In other words, this research has progressed from the organism level in one of two directions, either looking more closely at what's inside organisms ("skin-in biology") or looking more closely at how organisms interact with other organisms and their environments ("skin-out biology"). Third, theories of the origin and evolution of life are considered last because, to understand them, you need several ideas from the theories presented earlier. We begin with a look at the kinds of organisms that have been found on earth and present a theory of organism classification.

A THEORY OF ORGANISM CLASSIFICATION

Over 1.5 million kinds of organisms have been found on Earth, and more are discovered each year. Organisms can be compared to one another based on such observable features or properties as their length, weight, number of appendages, and color. These observable features are called characteristics or traits. A particular kind of organism is referred to as a *species*. **Species** are defined as groups of organisms that share enough characteristics that they can mate and produce fertile offspring.

A recent system of classification first divides all organisms into two major groups. The first group is made up of very simple cells with no distinct subcellular structures. These are called the **prokaryotes.** The second major group consists of organisms made up of cells that do have distinct subcellular structures, such as a nucleus. These are called the **eukaryotes.** Bacteria and blue-green algae are prokaryotes. All other organisms are eukaryotes.

The eukaryotes are further subdivided into four major groups: (1) those made of but a single cell (**protists**); (2) those that are multicellular and obtain their energy by eating other organisms (**animals**); (3) those that are multicellular and obtain their energy by absorbing food directly through their cell membranes (**fungi**); (4) those that are multicellular and able to use the energy of photons (light energy) to make their own food (**plants**). These five major groups—prokaryotes, protists, animals, fungi, and plants—are sometimes referred to as the five *kingdoms* of living things. Their relationships to one another are shown in Figure 7.1 on page 154.

Each of the five major kingdoms can be further divided into subgroups called *phyla* (the plural form of *phylum*) for animals and *divisions* for plants and plantlike protists. Phyla and divisions can be further divided into subgroups called *classes*. Classes can be divided into *orders*, orders into *families*, families into *genera*, and finally genera into *species*. Species are named by using a combination of the genus and species name. All biologists use Latin for naming organisms to avoid confusion that may arise when common names are used, which vary from country to country or even place to place within a country.

Testing Postulates of Theories of Organism Classification

One theory of organism classification argues that classification schemes should show the evolutionary development of the species being considered. Thus Figure 7.1 represents the evolutionary relationships among the major groups of organisms. Presumably, the more characteristics shared by two groups, the more closely they are related to each other and to a common ancestor. Both of these ideas illustrate a belief that organisms have evolved (i.e., changed across time). Obviously, the theory of evolution is different from the belief that organisms were created by an act of God "each according to its own kind." This

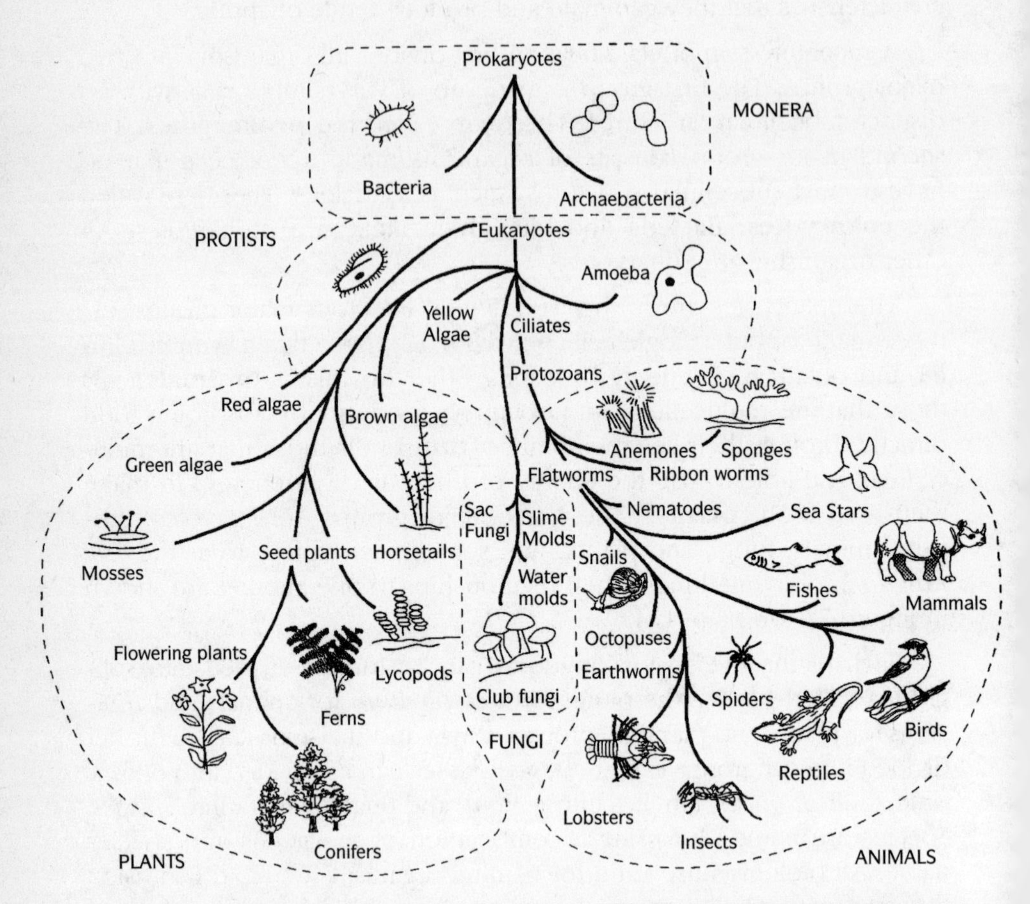

Figure 7.1 A modern five-kingdom biological classification showing possible evolutionary relationships. The animal kingdom, with its 1.3 million species, has the greatest number of species.

alternative belief, known as the theory of special creation, argues that God created specific kinds of organisms and that those kinds have changed little, if at all, since creation. What evidence can we get by observing species today will tell us which, if either, of these alternative beliefs is correct? As is always the case in testing ideas, we must first deduce the predicted consequences of these ideas. For the theory of special creation and the theory of evolution, the required reasoning is as follows:

Hypothesis: *If*...all the species of organism were created "each according to its own kind"...

Prediction: *then*...all organisms observed today should fit into a category (species), that depends upon the category into which they were first created.

On the other hand,

Hypothesis: *If*...species have changed and are still changing into different kinds from common ancestors...

Prediction: *then*...intermediate kinds of organisms (between two species) should exist.

Before we can turn to the evidence, we must first consider how one goes about deciding whether or not two organisms are of the same kind.

A Definition of Species

Biologists define a species as a particular type or kind of organism. But given a group of organisms, how can you tell if they are the same kind? Obviously, you look at them to see if they look alike or not. An ant is clearly a different kind of animal than an elephant. They do not look at all alike. But what about a red and a black ant? Are they two kinds of ants, two species, or merely one species with a variation in color?

To solve this problem, biologists have decided on a slightly different definition of species. This definition states that organisms are considered to be in the same species if they are enough alike so that they can mate and produce fertile offspring (that is, offspring capable of mating and producing offspring).

Let's consider a group of frogs that live in the eastern United States to see how this definition of species works. There are frogs that live in Vermont, others that live in New Jersey, and still others that live in Florida. All of these frogs look very similar, although their coloration is slightly different. According to the definition, to find out whether or not they are the same kind of frog (the same species), we must mate the frogs from different populations to see if they can produce fertile offspring.

When the frogs are mated, we discover that the Vermont frog and the New Jersey frogs can mate and produce fertile offspring. Also the

New Jersey frogs and the Florida frogs can mate and produce fertile offspring. This would lead us to conclude that the three populations of frogs are in the same species (Vermont frogs=New Jersey frogs=Florida frogs). But when we try to mate the Vermont frogs directly with the Florida frogs in one or the other environment, the frogs do not mate. This result suggests that the Vermont frogs and the Florida frogs are not in the same species!

One result says they are the same species, while the other result says they are different. Which is correct? The answer is that neither is correct. Instead, we have New Jersey frogs that are intermediate in kind to those in Vermont and Florida. This result then supports the hypothesis that the frogs are evolving and were not created in the form in which they are found today. Apparently all three frog populations had a common ancestor and have since changed into slightly different kinds, most likely as a consequence of different environmental conditions in Vermont, New Jersey, and Florida. Many other intermediate kinds of organisms have also been found, lending considerable support to the idea that kinds of organisms evolve.

POSTULATES OF THE MODERN THEORY OF ORGANISM CLASSIFICATION

1. A species is one kind of organism, defined as a group of organisms that share enough characteristics so that they are able to mate and produce fertile offspring.
2. Biological classification schemes represent the evolutionary development of the species considered.
3. The more characteristics shared by two groups, the more closely they are related to each other and the closer they are to a common ancestor.
4. Organisms can be subdivided into increasingly smaller and smaller groups of organisms with more and more characteristics in common. The larger groups are called *kingdoms,* and the smallest groups are called *species.*

THEORIES OF ORGANISM BEHAVIOR

Study of the behavior of living things is complex. It has led to numerous theories to account for various behaviors. However, biologists have identified a few dominant patterns of behavior, which are defined below. These dominant patterns of behavior vary considerably across different groups of organisms.

taxis (plural **taxes**) Simple, continuous movement either toward or away from a stimulus. The continuous movement of a planarian (flatworm) toward a light source is called a *positive photo taxis.* Taxes dominate in protozoa, simple metazoans, and worms, but play little or no role in vertebrates.

reflex Simple and essentially automatic response to a stimulus, such as the pulling away of your finger from a hot surface, or the sequence of orienting (locating) movements of a *Paramecium* away from a bubble of CO_2. Reflexes tend to dominate in the behavior of invertebraes. They are of less importance in vertebrates.

instinct Complex behavioral patterns acquired with little or no prior experience. Instincts tend to dominate in insects.

learning The acquisition of complex behavioral patterns primarily through experience. Learning begins to dominate in the lower vertebrates and dominates in primates.

reasoning The act of reflecting on prior experience to allow the modification of old behavior patterns or the invention of new ones. Reasoning is evident in some lower primates and dominates in humans.

It is important to note that no type of behavior, whether a simple taxis or a complex chain of reasoning, can develop without both genetic and environmental contributions. Thus the main difference among groups of organisms is not whether their behavior is controlled by genes or environment, but whether the organism has the *genetic potential* (natural ability) to acquire new behavioral patterns if it has the appropriate environmental stimulation. For example, a human infant has the genetic potential to speak. It responds to its environment (that is, other humans who speak to it) and thus acquires the ability to speak. A puppy does not.

Another important point about behavioral patterns is that they may vary from individual to individual within a given species. Some patterns lead to a greater chance of survival and reproduction; therefore, they tend to appear with increasing frequency in later generations. In other words, behavioral patterns may change (evolve) across time. (You will read more about the process of evolution in Chapter 13.)

MENDEL'S THEORY OF THE INHERITANCE OF CHARACTERISTICS

In a paper published in 1866, an Austrian monk named Gregor Mendel proposed a theory to explain why offspring tend to be similar to their parents. In other words, his theory was proposed to answer this question: How are the characteristics particular to a species transmitted from one generation to the next?

Mendel had conducted experiments with several types of pea plants with different characteristics. For example, in one experiment (shown in Figure 7.2), he started with some pea plants with purple flowers and others with white flowers. He collected pollen grains (which contain the male gametes) from the plants with the purple flowers and used them to pollinate (fertilize) the eggs (the female *gametes,* or sex cells) of the plants with the white flowers. He discovered that when the pea plants that resulted from this mating developed flowers, all of the flowers were purple.

Mendel then took pollen grains from some of these purple flowers and used them to pollinate the eggs of some of their sibling purple flowers (i.e., self-pollination). When the plants that resulted from this second mating developed flowers, 705 of the flowers were purple and 224 were white. In other words, the purple flowered plants outnumbered the white ones by a ratio of about 3 to 1. How could this result be explained?

Mendel's explanation was based on eight postulates. First, Mendel imagined that flower color is determined by tiny particles that he called *factors*. According to Mendel, every cell in the plants, including those in the flower, contain these factors. Thus, purple-colored flowers are produced by a specific purple factor and white-colored flowers are produced by a specific white factor. Mendel did not hypothesize what these factors looked like, nor did he try to explain how they determine flower color; nevertheless, he hypothesized that they existed.

Second, Mendel hypothesized that these factors actually are passed from parent to offspring. In his experiments, the sperm in the pollen

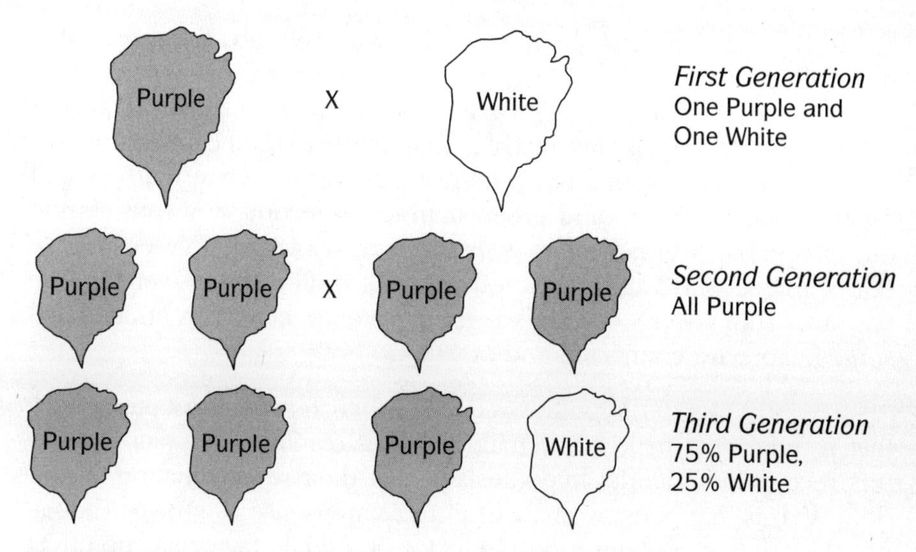

Figure 7.2 The results of Mendel's experiment with purple-flowered pea plants and white-flowered pea plants for three generations.

taken from one flower combined with the eggs from other flowers, so he assumed that the factors must be contained in the sperm and eggs (that is, the male and female sex cells, or gametes).

Third, Mendel hypothesized that each individual plant has a pair of factors for each characteristic in each of its cells. Half of the pair would come from the male parent by way of the sperm cell within the pollen and half from the female parent by way of the egg cell.

Fourth, Mendel hypothesized that when gametes are produced, pairs of factors are separated in such a way that each gamete receives one factor of each pair.

Fifth, Mendel hypothesized that when a particular gamete is being produced by the plant, there is an equal (50-50) chance that the gamete will receive one or the other factors of a pair. In other words, there is a 50% chance that any particular cell will receive a purple factor and a 50% chance that it will receive a white factor.

Sixth, when a gamete is receiving a factor for flower color, it is also receiving factors for other characteristics such as leaf shape, pod color, and plant height. Mendel hypothesized that this distribution of factors to the gamete occurs independently. In other words, if a gamete receives a purple factor for flower color and is about to receive a factor for pod color (for example, yellow or green), the fact that it has already received a factor for purple flower color will have no effect on which pod color factor that it receives. That is, the gamete is just as likely to receive a yellow factor as it is a green factor. Here it may be helpful to think of the factors as tiny marbles that are not connected to one another. By analogy, if one has a bag of purple and white marbles and another bag of yellow and green marbles, selecting a purple marble out of one bag will not affect your chances of picking a green marble out of the second bag. On the other hand, if all marbles were in one bag and some were connected, things would be different. Perhaps some factors are connected and others are not!

Seventh, Mendel hypothesized that when fertilization takes place, that is, when the male and female gametes combine, the pairs of factors recombinc. Mendel hypothesized that their recombination is random (that is, happens by chance) in the same sense as above. That is, whether an egg contains a purple factor or a white factor has no effect on the type of factor (purple or white) it will receive from the sperm.

Finally, Mendel hypothesized that one member of a pair of factors sometimes, somehow, exerts more influence than the other factor, so that its characteristic shows up in the plant. For example, a purple flower factor paired with a white flower factor can exert its influence so that the flower will turn out purple and not some intermediate color like pink. Although Mendel did not hypothesize how this occurs, he called it *dominance*. In this case, the purple factor is the *dominant* factor, and the white factor is the *recessive* factor.

Mendel's postulates taken together can be used to explain the results of his experiment with purple and white flowers, as shown in Figure 7.3. Before you read the explanation below, see if you can use Mendel's postulates to explain the results.

As shown in Figure 7.3, the purple flower male has cells with a pair of factors designated P P, and the white female flower has cells with a pair of factors designated W W.[1] When gametes are produced, the male gamete cells receive one or the other of the P purple factors, so, in terms of its factor for flower color, all male gametes are the same. Likewise, when female gametes are produced in the female flower, all will

---

[1]This is nonstandard use of capital letters to designated pairs of genes (alleles). I am using it because it is easier for a beginner to understand. Standard practice uses capital and lower-case letters to designate dominant and recessive alleles (see postulate 8 on page 162).

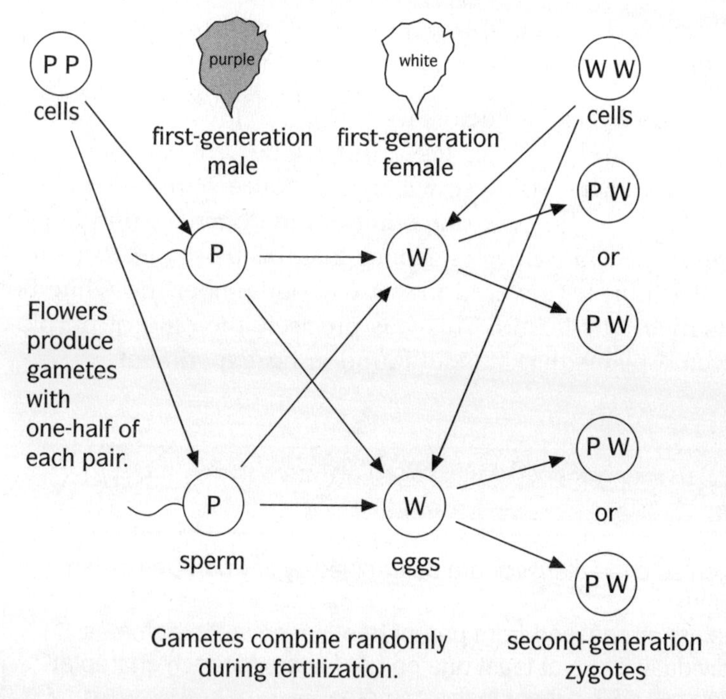

Figure 7.3 Hypothesized transfer of factors from first- to second-generation plants to produce second generation offspring with purple flowers.

receive one or the other of the white factors, so all female gametes contain one white factor.

When the male and female gametes combine during fertilization, all possible combinations of gametes produce **zygotes** (that is, the first cell of the next generation that will divide repeatedly to produce a new individual) with a purple factor from the male and a white factor from the female. So they all are P W and all of the plant cells that develop from these zygotes will also contain P W. If we assume that the purple factor dominates the white factor, then all the flowers in this second generation will be purple.

Now suppose that the P W plants produce gametes. According to the theory, the purple and white factors separate, so that two types of male and two types of female gametes are produced. The male will be either P or W. Likewise, the female will be either P or W.

When fertilization takes place randomly, as shown in Figure 7.4, four types of zygotes can be formed: P P, P W, W P, W W.

The P P zygote has two factors for purple, so it will produce purple flowers. The W P and P W zygotes both have a purple factor that is assumed to be dominant, so these will also produce purple flowers. The fourth type, W W, is the only one without the dominant purple factor, so it will produce white flowers. Notice that this theory leads us to the prediction that purple-flowered plants will outnumber the white-flowered plants by a 3 to 1 ratio. This was precisely the ratio of purple- to white-flowered plants that Mendel found in his experiment.

POSTULATES OF MENDEL'S THEORY OF INHERITANCE

1. Inherited characteristics are determined by tiny particles called factors.
2. Factors are passed from parent to offspring in the gametes.
3. Individuals have at least one pair of factors for each characteristic in all cells except in the gametes.
4. During gamete formation, paired factors separate. Each gamete receives one factor of each pair.
5. There is an equal chance that a gamete will receive either one of the factors of a pair.
6. With two or more pairs of factors, the factors of each pair assort independently to the gametes.
7. Factors of a pair that are separated in the gametes recombine randomly during fertilization.
8. Sometimes one factor of a pair dominates the other factor, so that it alone controls the characteristic (dominant/recessive).

Today Mendel's "factors" are called **genes**. The genes are made up of very complex molecules arranged next to one another in a long row. A long row of genes is called a **chromosome**. Since genes occur in pairs, so do chromosomes. Pairs of genes that occur at the same location of a pair of chromosomes and function to dictate a specific characteristic (such as flower color) are called **alleles**. The actual combination of observable characteristics of an individual is called its

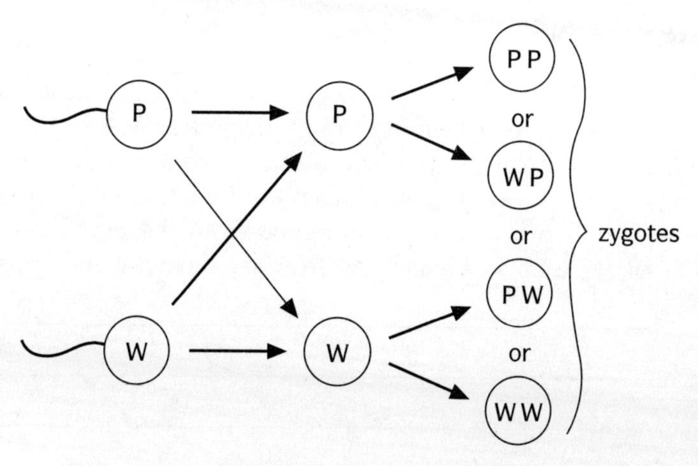

Figure 7.4 Fertilization of third-generation zygotes.

phenotype. The kinds of genes that are hypothesized to exist to dictate the phenotype is called its **genotype**.

Mendel's theory explains how organisms are able to produce offspring similar to themselves. Occasionally, however, characteristics show up in offspring that do not appear to be the result of Mendelian processes. Instead, they appear to be the result of rearrangements in the genes themselves. These rearrangements, called **mutations**, are often harmful, even fatal to organisms. But occasionally mutations are helpful by producing a characteristic that gives the organism a competitive advantage over other members of its species.

QUESTIONS

1. The evolution of species can also be considered to be an example of the emergence of new properties. For example, compare bacteria with the protista. Name a property that emerged during the evolution of protists from their bacteria-like ancestors. Name several properties that emerged from the evolution of a higher plant, such as a pine tree, from its algaelike ancestors.

____

____

____

____

2. Although the presence of many intermediate forms of living things clearly supports the idea that living things are changing (evolving) through time, is it not possible that an all-knowing, all-powerful creator could have created all of these intermediate forms? Assuming that your answer to this question is yes, what does this tell us about our ability to scientifically test explanations that involve all-powerful entities?

____

____

____

____

3. Explain why the use of a control group is common practice when conducting scientific experiments; that is, what would be the consequences of not having a control group?

____

____

____

____

4. As discussed in Chapter 3, salmon are able to navigate long distances. Adult salmon navigate to precisely the same stream in which they were born to lay their eggs. In addition to the sight,

smell, and magnetic field hypotheses, generate a fourth hypothesis. Design an experiment to test it. Have you designed a controlled experiment? Explain.

5. Of course, not all characteristics occur in 3 to 1 ratios as described previously. Suppose, for example, the ratio of flower colors in Mendel's experiments had turned out to be one purple to two pink to one white? How could Mendel's postulates be modified or extended to explain this ratio? How could you modify the postulates to explain the 1 to 1 ratio of males to females in most species of organisms? Also, what sort of ratio of offspring characteristics would you expect if the characteristics in question were controlled by more than one pair of genes (e.g., two pairs, three pairs)?

6. Notice that postulate 6 of Mendel's theory states that when considering two or more pairs of factors, the factors of each pair assort independently to the gametes. In other words, if one pair of genes controls eye color (say blue or green) and another pair of genes controls hair color (say brown or blond), then this postulate claims that whether you get one or the other gene for eye color (i.e., blue or green) will have no effect on which gene you get for hair color (i.e., brown or blond). In other words, in this example eye color and hair color are not linked. They assort (travel) independently. But it is known that one's sex (male or female) is controlled by genes and so is whether or not one becomes color-blind. It is also known that the male gene and the gene for color blindness are linked. In other words, typically it is the male that becomes color-blind, not the female. Does this fact

contradict postulate 6? If so, should we, therefore, reject Mendel's theory? Explain.

7. Many people think that one's intelligence is controlled primarily by one's genes. Others claim that one's intelligence is controlled primarily by the environment. What sort of evidence would be needed to test these ideas?

TERMS TO KNOW

allele
animal
characteristic
chromosome
dominant (allele)
eukaryote
fungus, pl. fungi
gene
genotype
instincts
learning
mutation
phenotype
plant
prokaryote
protist
reasoning
recessive
reflex
species
taxis
zygote

SAMPLE EXAM

Circle the best answer for each question.

1. Which category of classification contains organisms that are most closely related?

    a. family

    b. class

    c. order

    d. genus

    e. kingdom

The next three questions are based on Table 7.1. (Hint: Fill in the blank spaces *before* answering the questions).

TABLE 7.1

	ORGANISM I	ORGANISM II	ORGANISM III	ORGANISM IV
Phylum	Arthropoda			
Class	Hexopoda			
Order	Lepidoptera	Lepidoptera		
Family	Tortricidae	Psychidae		Tortricidae
Genus	Archips	Solenobia	Archips	Eulia
Species	rosana	walshella	fervidana	pinatubana

2. Which two organisms are most closely related?

    a. I and II

    b. I and III

    c. I and IV

    d. II and III

    e. II and IV

    f. III and IV

3. Which organism is the most distantly related to organism IV?

   a. I

   b. II

   c. III

   d. All are equally related.

4. Which organisms belong to the class Hexapoda?

   a. I and III

   b. I and IV

   c. I, II, and IV

   d. I and IV

   e. I, II, III, and IV

5. If a cross were made between two black, rough-haired guinea pigs and the resultant offspring included six with black rough hair and one with white rough hair, from this information it can be assumed that

   a. one of the parents carried genes for white hair.

   b. both the parents carried genes for white hair.

   c. a mutation had occurred.

   d. white hair is a dominant trait.

6. "A" represents the gene for a dominant trait and "a" its recessive allele. If Aa mates with aa,

   a. all offspring will be of the dominant phenotype.

   b. all offspring will be of the recessive phenotype.

   c. 50% of the offspring will be of the recessive phenotype.

   d. 75% of the offspring will be of the dominant phenotype.

7. What is the probability that a parent with a Ddee genotype will produce a gamete with a de genotype?

   a. 0%

   b. 25%

   c. 50%

   d. 75%

   e. 100%

8. What is the probability that the mating of a male with a DdEe genotype and a female with a DdEE genotype will produce an offspring with a DdEe genotype?

   a. 0%

   b. 25%

   c. 50%

   d. 75%

   e. 100%

CHAPTER 8

ORGAN AND SYSTEM LEVEL THEORIES

GETTING FOCUSED

- *Learn about Harvey's theory of circulation.*
- *Learn about a theory of heart rate regulation.*
- *Learn about a theory of water rise in plants.*
- *Evaluate how hypotheses about physiological processes are tested.*

HARVEY'S THEORY OF BLOOD CIRCULATION

Prior to the thinking and research of a seventeenth-century English scientist named William Harvey, most people believed that blood was manufactured in the heart and was consumed in various parts of the body. However, in 1628 Harvey argued against this view in a book called *On the Motion of the Heart and Blood in Animals*. In that book Harvey proposed a theory of blood circulation that consisted of these six basic postulates:

POSTULATES OF HARVEY'S THEORY OF BLOOD CIRCULATION

1. Blood flows continuously in one direction from the heart and back to the heart.
2. Blood passes from the right ventricle of the heart to the lungs, back to the left auricle, and from there to the left ventricle.
3. From the left ventricle, blood is forced into the aorta and through branches of arteries to all parts of the body except the lungs.
4. From the smallest branches of the arteries, the blood flows into tiny capillaries into the smallest veins.
5. From the smaller veins, the blood flows into larger and larger veins and then into the right auricle of the heart.
6. The right auricle forces the blood periodically into the right ventricle.

Harvey argued against the old theory of blood manufacture and consumption and for his theory of circulation. He estimated that the right ventricle holds about two ounces of blood. He then reasoned that if the heart pumps 72 times each minute, then in 60 minutes the heart will pump no less than $2 \times 72 \times 60 = 8{,}640$ ounces = 540 pounds of blood, three times the weight of a heavy man! Because it is not reasonable to think that this much blood is manufactured or used in one hour, this reasoning supports the postulate that blood circulates.

Harvey's theory of circulation also leads to an interesting prediction that can be used to further test the theory. It was known at the time that if a person handles a lot of garlic, his or her breath will soon smell garlicky. This observation suggests that garlic molecules pass through the skin into the blood and then circulate, eventually ending up in the mouth. To test this idea, a tall man with bare feet was placed horizontally through a hole in a wall, his feet in one room and his head in another. Garlic was then rubbed on the bottom of his feet.

The reasoning used to test the circulation postulate is as follows:

Hypothesis: *If...* blood circulates from the feet to the mouth...

and... garlic is rubbed on the feet and enters the blood...

Prediction: *then...* the garlic smell should be on the man's breath.

On the other hand...

Hypothesis: *If...* blood is consumed by the feet and does not circulate...

Prediction: *then...* the garlic molecules should not be on the man's breath.

The result of the experiment was that in just a few seconds, an observer standing by the man's head smelled garlic on his breath. Since this is the result predicted by the circulation theory, and not by the blood manufacture and consumption theory, the circulation theory is supported. Of course, many other experiments and observations have been made since Harvey's time that also support the circulation theory.

A THEORY OF HEART RATE REGULATION

Today Harvey's theory of blood circulation is well accepted. Blood circulates throughout the body through arteries and veins due to the pumping action of the heart. The blood's primary function is to carry needed food and oxygen molecules to the body cells and to carry away carbon dioxide and poisonous waste molecules, such as urea. The rate at which molecules are delivered to and transported away from the cells depends upon the rate at which the heart pumps. Therefore, the regulation of heart rate is extremely important to the body's cells. If the rate is too low, the cells will not receive enough food and/or oxygen molecules to sustain their level of activity. Also, waste molecules will accumulate and may kill the cells. A too rapid heart rate is also harmful as too much energy must be used to sustain the high rate and the heart and/or the blood vessels may be damaged.

Exactly how heart rate is maintained at just the right level is an interesting question. For example, it has been shown that exercise increases heart rate. Let us consider a modern-day theory of how this may happen. Increased muscle cell activity presumably increases the amount of carbon dioxide (CO_2) produced in the cells. This increased amount of CO_2 passes into the blood. Here it combines with water (H_2O) to produce H_2CO_3 molecules called carbonic acid. The carbonic acid molecules then split to form hydrogen ions (H^+) and bicarbonate ions (HCO_3^-). The increased number of hydrogen ions in the blood is then sensed by special cells in the aortic arch and in the carotid sinus that send electrical signals to a location in the brain stem (called the cardio-accelerator center). The center then relays signals to the heart muscle, which cause it to beat more rapidly.

The resulting increase in heart rate will then pump more blood to the lungs, allowing the CO_2 to leave the blood via the lungs. A drop in the CO_2 will then indirectly cause a drop in the number of hydrogen ions, which will have the reverse effect (via a cardio-inhibitory center in the brain stem) and reduce the heat rate. This series of events regulates heart rate, keeps the cells supplied with the necessary numbers and kinds of molecules, and insures that waste molecules are transplanted from the cells. It is an excellent example of **homeostasis,** that is, the ability of living things to maintain nearly constant internal conditions (see Figure 8.1).

Rebreathing air also has been found to increase heart rate. Because rebreathing air would also increase the CO_2 level in the blood (rather than in the lungs, where it could escape), the same mechanism could account for the increase in heart rate.

One possible explanation for a decrease in heart rate with decreasing temperature (e.g., putting an arm into an ice bath) goes something like this: Lower temperature causes a constriction of blood vessels in the arm (a response to cold temperatures to keep core body temperature up). Constriction of blood vessels increases blood pressure, which is sensed in the carotid sinus and aortic arch. Signals are then sent to the brain, which returns inhibitory signals (signals that interfere with an action) to the heart to decrease heart rate, which lowers blood pressure to its original level.

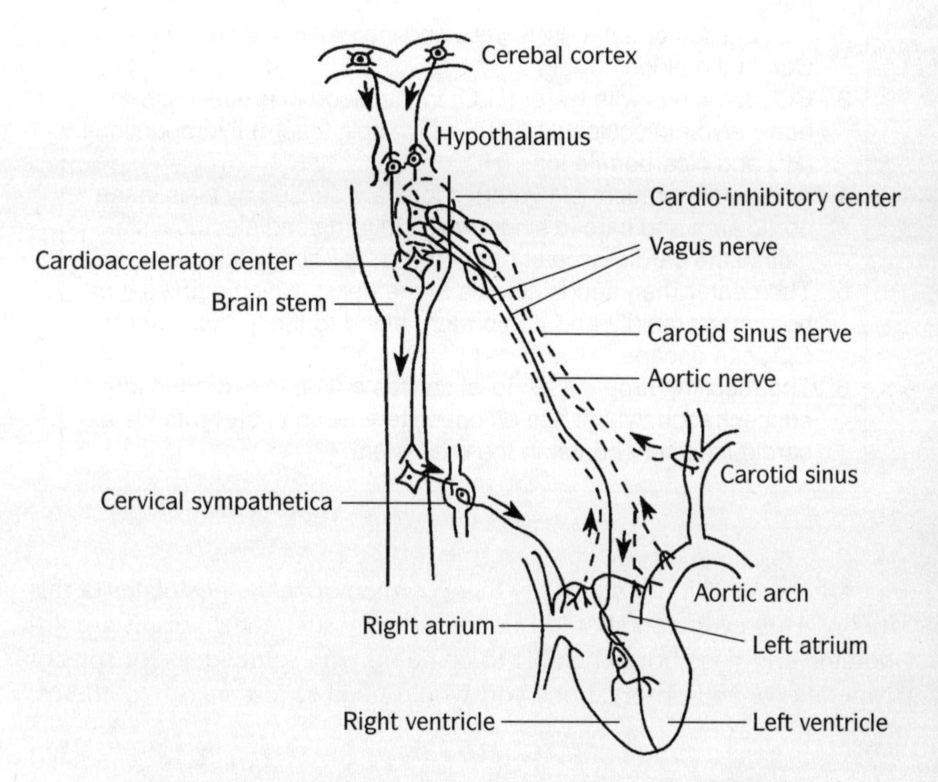

Figure 8.1 Structures involved in the regulation of heart rate.

As you might guess, many different theories have been proposed to explain how the body regulates itself to maintain homeostasis. The following postulates summarize the theory of muscle cell activity and heart rate:

POSTULATES OF THEORY OF MUSCLE CELL ACTIVITY AND HEART RATE

1. Increased muscle cell activity increases the amount of CO_2 in the cells.
2. CO_2 passes out of muscle cells and increases the amount of CO_2 in the blood stream.
3. CO_2 combines with water (H_2O) in the blood to produce carbonic acid molecules (H_2CO_3), which split to form hydrogen ions (H^+) and bicarbonate ions (HCO_3^-).
4. Increased numbers of hydrogen ions are sensed by cells in the aortic arch and carotid sinus, which in turn send electrical signals to the cardio-accelerator center in the brain stem.
5. The center then sends signals to the heart, which causes it to beat more rapidly and pump more blood to the lungs, where CO_2 can escape.
6. The resulting drop in CO_2 level causes a drop in hydrogen ion concentration, which has an opposite effect on heart rate via a cardio-inhibitory center in the brain stem.

Of course, is it not necessary that you memorize the postulates of this theory. However, you should understand why such mechanisms are important, and you should be able to come up with some ideas for sorts of experiments and observations you would need to test such hypotheses.

A THEORY OF WATER RISE IN PLANTS

The stems of vascular plants contain vessels called **xylem** that conduct water from the roots to the leaves, where it is used for photosynthesis and other vital cell processes. But what causes water to rise against the physical force of gravity? Apparently, a number of mechanisms are involved.

One force results from the osmotic movement of water (see Chapter 10) into roots from the soil. This osmotic force, called **root pressure,** is generated at the bottom of the xylem and tends to push water upward. As evidence of root pressure, cut stems "bleed" fluid for some time after being cut. Root pressure is also presumably responsible for the occasional appearance of drops of water on the tips of leaves at the leaf vein endings. This occurs when water loss due to evaporation (called **transpiration**) is low and when the soil contains a lot of water, a process called **guttation.** Root pressure alone, however, is not strong enough to account for the movement of water up a tall tree. Another force or set of forces must be involved. One of these forces appears to be the cohesion of water molecules. Water molecules tend to stick to each other because they are *polar.* They may also stick to the walls of the xylem tubes. A **polar molecule** is one in which the negatively charged electrons do not distribute themselves equally, so one part of the molecule ends up with a net negative charge and another part ends up with a net positive charge. Because the oxygen atom ends up with a slight negative charge and the hydrogen atom ends up with a slight positive charge, they are attracted to each other, and a group of water molecules stick together. Thus, a pull on the top molecules will result in the rise of the entire column.

But what sort of a pull can exist at the top of the column? One hypothesis suggests that the evaporation of water from the leaves will cause a partial vacuum that can suck water up, like drinking a milkshake through a straw. Clearly, however, this cannot be the case as suction as a force is nonexistent. The force that moves the milkshake up the straw is a push from below due to greater air pressure on the surface of the milkshake outside the straw than on the surface inside the straw (see Kinetic-Molecular Theory in Chapter 10).

What, then, provides the pull? A possible explanation at this point involves osmosis (see A Theory of Molecular Movement in Chapter 10) and goes as follows: Evaporation of water from leaf cells increases the concentration of large molecules in the leaf cells, therefore increases the osmotic pull of water molecules into the leaf cells from nearby cells, such as those in nearby xylem tubes. Because water is cohesive, this osmotic pull at the top will cause the entire column of water in the xylem to rise.

Although this explanation has gained some acceptance among plant physiologists, it leaves a few problems unresolved. The explanation requires the maintenance of the column of water in the xylem; yet breaks occur frequently. How the explanation can allow for this contradictory finding is not clear. Another puzzle is how the column of water is established in the first place. Perhaps it "grows" there as the plant grows.

POSTULATES OF THE WATER RISE THEORY

1. Root pressure due to osmosis pushes water part of the way up xylem tubes in stems.
2. Water molecules are polar; therefore, they are cohesive.
3. Evaporation of water from leaf cells (transpiration) increases the concentration of large molecules in the leaf cells, therefore increasing the osmotic pull of water molecules from nearby xylem tubes.
4. Osmotic pull in leaf cells near the top of a column of water in a xylem tube will cause the entire column to rise.

QUESTIONS

1. Postulate 2 of the theory of heart-rate regulation proposes that elevated levels of CO_2 in the blood stream are caused by increased muscle cell activity. Propose two other possible causes of elevated levels of CO_2 in the blood. How could you experiment to determine which if any or all of the three possible causes is the actual cause?

   ___

   ___

2. Postulates 4 and 5 of that theory refer to the activity of a hypothesized cardio-accelerator center in the brain. Assuming that this center can be located, describe an experiment to test its existence and its hypothesized function. Use the if…and…then… pattern of thinking in your description.

   ___

   ___

   ___

3. In vascular plants, xylem tubes conduct water from the roots through the stem to the leaves where the water is used for photosynthesis. Photosynthesis results in the production of food and oxygen molecules. Some of these food molecules may pass

down the stem to the roots through special tubes called **phloem.** The roots store excess food until it is needed by the plant. Propose at least three alternative hypotheses to explain how the food molecules pass through the phloem tubes to reach the roots; that is, what force or forces might cause the food molecules to move down the phloem tubes?

___

___

___

4. How could you experiment to test these alternatives?

___

___

___

TERMS TO KNOW

guttation
homeostasis
phloem
polar molecule
root pressure
transpiration
xylem

SAMPLE EXAM

The first four questions are based on the graph and key in Figure 8.2. Use the data on the graph to classify the statements.

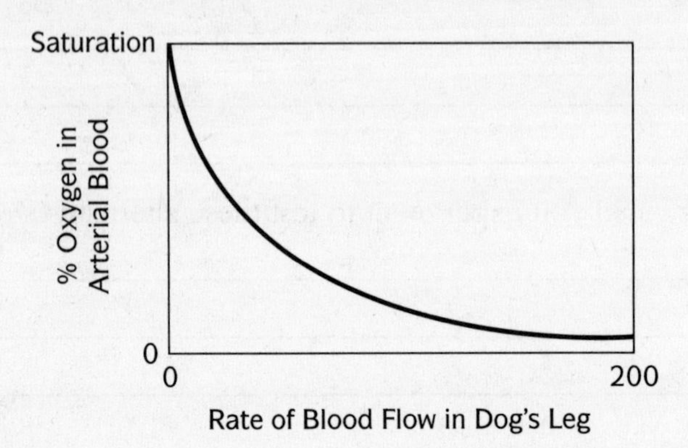

Figure 8.2 The rate of blood flow and oxygen saturation in a dog's leg.

KEY: a. A hypothesis consistent with the results.
b. A hypothesis contradicted by the results.
c. A statement of the results.

1. There is a correlation between the oxygen level in the blood and the rate of blood flow.

 ____

2. Oxygen lack causes contraction of blood vessels and thus reduces blood flow.

 ____

3. Oxygen lack causes an increase in size in blood vessels.

 ____

4. Greater amounts of oxygen of the blood bring about a marked increase in blood flow by increasing heart rate.

 ____

The two hypotheses below are alternative explanations of the control of pancreatic secretion into the intestine.

I. Nerves stimulate the pancreas to secrete its enzymes into the intestine.

II. A hormone in the blood causes the pancreas to secrete its enzymes into the intestine.

KEY:
a. Supports hypothesis I only
b. Supports hypothesis II only
c. Supports both hypotheses

Use the key above to classify each of the following experiments (5–7) as they relate to the hypotheses.

5. The pancreas is stimulated when food enters the small intestine of a normal animal. Food in the intestine could stimulate a nerve that could stimulate the pancreas. Likewise, food in the intestine could stimulate the production of a hormone that could pass into the blood to stimulate the pancreas. Because both of these possibilities are consistent with the observation, they both are supported by the observation.

____

6. When a nerve leading to the pancreas is stimulated, the pancreas secretes enzymes.

____

7. When the nerves leading to the pancreas are cut and weak acid is placed into the intestine, the pancreas secretes enzymes.

____

8. An animal that attaches to the side of a fish and sucks the fish's blood for food would probably have

a. a long digestive tract with many blind pouches.

b. a short, simple digestive tract.

c. no digestive tract.

d. a digestive tract unlike that of any known animal.

9. The fact that the small intestine has an inner surface with many folds and is supplied with a rich supply of blood suggests that an important function is to

   a. produce chemicals used in digestion.

   b. support the body.

   c. absorb food into the blood stream.

   d. transfer food from the stomach to the large intestine.

   e. receive signals from the brain to coordinate digestion.

CHAPTER 9

CELL LEVEL THEORIES

GETTING FOCUSED

- *Evaluate a theory of cell replication.*
- *Compare the theories of sexual reproduction and meiosis.*
- *Learn about a modern theory of embryological development.*
- *Learn how cellular level hypotheses are tested.*

CELL THEORY

The Greek philosopher/scientist Aristotle (384–322 B.C.) knew quite a lot about the organs and organ systems that make up complex living things. But he and others in ancient Greece knew nothing about the cells and tissues that make up the organs, because they did not have microscopes. The first microscopes were not invented until the 1600s.

In 1665 Robert Hooke, an English scientist, used one of the early microscopes to observe a thin slice of bark. He coined the term *cells* to name the tiny rectangular compartments that he saw. Today we know that Hooke was observing plant cell walls. Although Hooke's observations encouraged others to use microscopes to examine the structure of living things, it was not until 1824 when the French scientist Dutrochet (doo-trah-SHAY), after comparing a variety of plant and animal tissues, proposed that *all* tissues are composed of cells. This generalization was supported by the observations of two German scientists, Theodor Schwann and Matthias Schleiden. In an 1838 paper, they reported their observations of a variety of plant and animal cells and proposed that cells are not only the basic structural units of living things, but they are the basic functional units as well.

Another important part of cell theory was first proposed in 1855 by the German biologist, Rudolf Virchow (FIR-koh) when he hypothesized that new cells arise by the division of existing cells. Of course, this hypothesis is consistent with the theory of biogenesis (see Chapter 6).

Classic cell theory consists of three basic postulates.

POSTULATES OF CELL THEORY

1. Cells, which in general consist of an *outer cell membrane, cytoplasm,* and a *nucleus,* are the structural and functional units of all organisms.
2. Cells arise only from preexisting cells.
3. Cells contain hereditary material in the nucleus by which specific characteristics are passed from parent cells to daughter cells.

This third postulate is a relatively recent addition to the cell theory. The hereditary material, called *genes,* consist of *d*eoxyribo*n*ucleic *a*cid (DNA) molecules that are organized in precise sequences and are passed from one cell to the next to ensure the continuity of characteristics from one generation to the next. Mendel referred to the hereditary material as "factors." (See Chapter 7. See also Chapter 11 on the structure and function of DNA.) In addition to the basic postulates of the cell theory stated above, a variety of subtheories about cell structure and function exist that attempt to explain how cells carry out specific activities, such as metabolism, absorption, secretion, movement, growth, and division.

Discovering the Importance of the Cell Nucleus

A prominent structure that can be easily observed near the center of many cells is the **nucleus.** One of the important early observations of the role of the nucleus was made over 100 years ago by a German embryologist, Oscar Hertwig. Hertwig was observing the egg and sperm cells of sea urchins and discovered that when a sea urchin's egg is fertilized by a single sperm, the sperm's nucleus is released and fuses (joins) with the nucleus of the egg cell. This observation provided support for the hypothesis that the nucleus carries the genetic information from one generation to the next, because it appeared to be the only material passed from father to offspring in the nucleus of the sperm.

Following Hertwig's observations, a number of experiments were performed to further investigate the role of the nucleus. In one experiment, the nucleus of an amoeba was removed. The amoeba stopped dividing and died within a few days. However, when a nucleus from another amoeba was injected into such an amoeba, it was able to survive and continue dividing. Therefore, it appears that the nucleus is necessary for the continued life and division of cells.

In the early 1930s, Joachim Hammerling investigated the function of the nucleus and the cytoplasm in a marine algae called *Acetabularia. Acetabularia* consists of one huge cell two to five centimeters in height. The cells consist of a cap, a stalk, and a foot. The foot contains the cell's nucleus (see Figure 9.1).

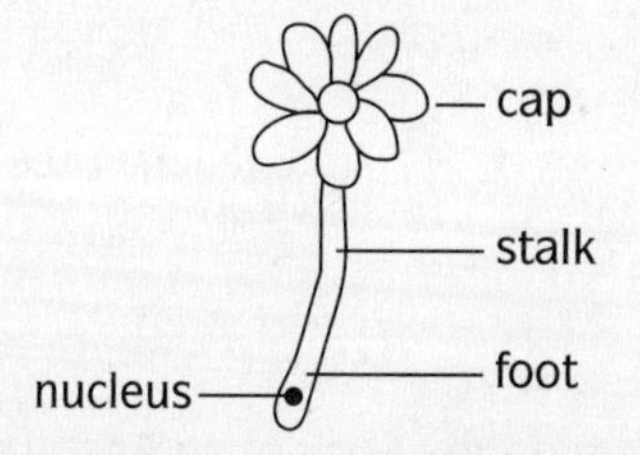

Figure 9.1 An *Acetabularia* cell.

In one of Hammerling's experiments, he explored the role of the nucleus in the regeneration of cell parts that had been cut off. Hammerling hypothesized that the nucleus controls regeneration, and he reasoned as follows:

Hypothesis: *If*...the nucleus controls regeneration...

and...the stalks of several cells are cut...

Prediction: *then*... the bottom sections that contain the foot and nucleus should be able to generate new caps but the top section should not be able to generate a new foot and nucleus.

Because the experiment turned out just as predicted, Hammerling's hypothesis was supported (see Figure 9.2). Apparently, the nucleus controls the regeneration of cell parts.

In a subsequent experiment, Hammerling cut the stalks of other cells in two places instead of one (see Figure 9.3). Based upon his prior thinking and results, he came up with a similar prediction:

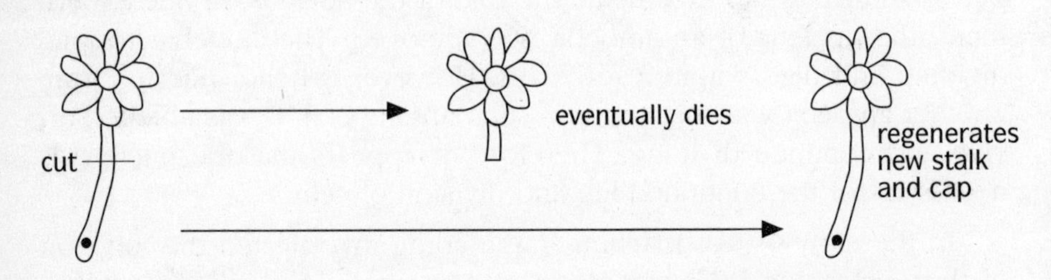

Figure 9.2 The result of cutting an *Acetabularia* cell in half.

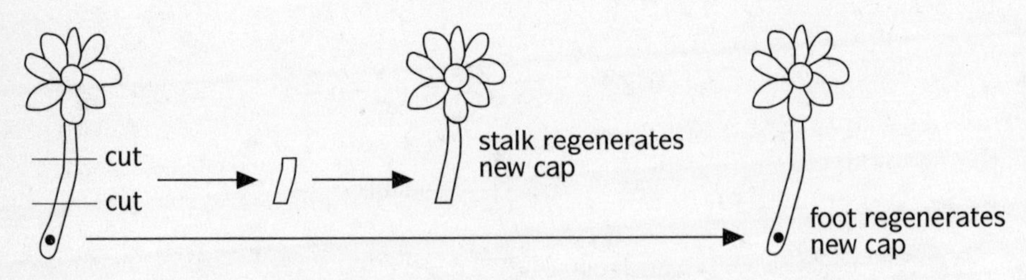

Figure 9.3 The result of cutting the stalk of an *Acetabularia* cell in two places.

Hypothesis: *If*...the nucleus controls regeneration...

and...the stalks are cut in two places...

Prediction: *then*...only the section that contains the foot and nucleus should be able to generate new stalks and caps.

The result of this experiment, shown in Figure 9.3, was not as predicted, as both the foot and the control stalk were able to regenerate new caps. Hammerling was, therefore, prompted to modify his hypothesis. Perhaps the nucleus controls regeneration of all parts by producing molecules that pass into the cytoplasm. Thus, if the cytoplasm contains enough of these molecules, regeneration can take place even when the nucleus has been removed.

To test this idea, Hammerling cut the newly regenerated caps off the stalks of the footless cells and waited to see if the stalks were able to generate caps again. Hammerling's molecule-production hypothesis led to the prediction that the stalks should *not* be able to regenerate caps again. Hammerling suspected that the necessary molecules would have been used up when the first caps were generated and no nucleus was present to produce more molecules. This experiment turned out as predicted, suggesting that the nucleus exerts its control by somehow producing molecules that pass into the cytoplasm. This, and other experiments of this sort, have led biologists to conclude that the nucleus controls virtually all activities of the cell by producing several different kinds of molecules. Modern research is focused on determining just what those molecules are, how they are produced, and how they exert their control.

A THEORY OF CELL REPLICATION

In multicellular organisms, growth results from the addition of new cells or the increase in size of existing cells. The addition of new cells is accomplished by a process in which an existing cell duplicates its parts and then separates the parts into two sets. The sets then migrate to opposite halves of the cell, at which time a new cell membrane—and in plants a new cell wall and a new membrane—forms between the two halves. The result is two new cells, each with a complete set of parts. The entire process is called **cell replication.**

The replication and migration of the cell parts found in the cell nucleus (i.e., the chromosomes) is called **mitosis.** The replication and migration of the cell parts found *outside* the nucleus (e.g., the mitochondria, chloroplasts) is called **cytokinesis.**

The chromosomes direct the cell's activities. Each chromosome—indeed each segment of each chromosome—directs a specific cellular activity; therefore, it is imperative that each new cell produced by cell replication receives a complete set of chromosomes. The process of mitosis enables this to happen. For the sake of clarity, mitosis is described and diagrammed for only a single chromosome; however, you should keep in mind that all of the other chromosomes are undergoing the same processes.

Mitosis begins when the uncoiled chromosome (see Figure 9.4A) replicates itself (Chapter 11 discusses this process). This produces two identical chromosomes, which now begin to coil up and become clearly visible in a light microscope (see Figure 9.4B). The coiling makes the chromosomes much less likely to get tangled as they migrate. However, coiled chromosomes are unable to direct the cell's activities. The nuclear membrane begins to disintegrate at this time.

Next, the replicated chromosomes migrate toward the center of the nucleus. When they arrive at the center, they line up (as do all of the other replicated chromosomes) at what is referred to as the *equatorial*

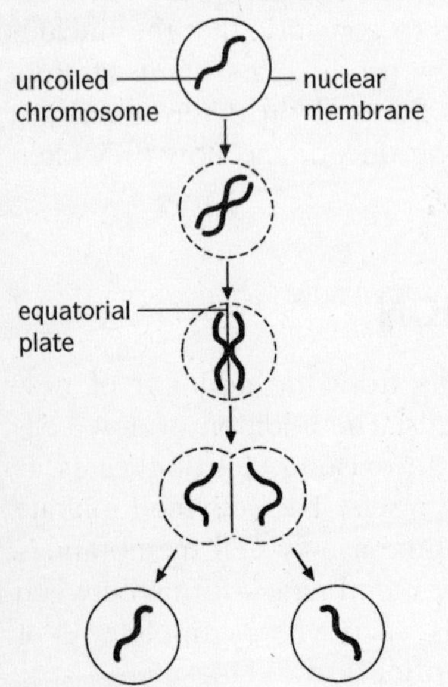

A. Uncoiled chromosomes direct the cell's activities (only one chromosome is shown).

B. Chromosome replicates itself. The two resulting chromosomes coil up and remain attached at their midpoints. Nuclear membrane begins to disintegrate.

C. Replicated chromosomes migrate to center and line up on either side of the equatorial plate.

D. Replicated chromosomes split apart and migrate to opposite sides of the cell. They bend at midpoint as they move. Cytokinesis begins.

E. A new nuclear membrane forms around each set of chromosomes. A new cell membrane also forms and cytokinesis is completed.

Figure 9.4 The process of mitosis as depicted for a single replicating chromosome.

plate. One of the replicated chromosomes lines up on the right side of the plate, and the other lines up on the left side (Figure 9.4C).

The replicated chromosomes then split apart and begin moving to opposite sides of the cell. As they move, they bend at the midpoint (Figure 9.4D). In most cells, cytokinesis begins when the chromosomes reach the opposite sides of the cell.

Last, a nuclear membrane forms around each new set of chromosomes (Figure 9.4E). The cell splits in two with the development of a new cell membrane—and in plants a new cell wall—near the old equatorial plate. Cytokinesis is completed at this time.

POSTULATES OF CELL REPLICATION THEORY

1. Cell replication involves the precise replication of cell parts inside the nucleus (mitosis) and outside the nucleus (cytokinesis) in a continuous sequence.
2. First, the very long thin chromosomes replicate themselves.
3. The replicated chromosomes then coil up, and the nuclear membrane disintegrates.
4. Next, the replicated, coiled chromosomes line up at the equatorial plate.
5. The replicated, coiled chromosomes separate into two identical groups and migrate to opposite sides of the cell. Cytokinesis begins in cell parts outside the nucleus.
6. Finally, the chromosomes arrive at opposite sides of the cell and uncoil. Cytokinesis is completed, and a nuclear membrane forms around each set of uncoiled chromosomes. The cell divides in two with the formation of a new cell membrane, and in plants a cell wall along the old equatorial plate.

Testing a Postulate About Chromosome Movement

The fifth postulate of the cell replication theory states that the coiled chromosomes migrate to opposite sides of the cell. Time-lapse photography of dividing cells leaves no doubt that this is the case. But what actually causes the chromosomes to move? Microscopic examination of migrating chromosomes shows that they are attached at their centers (**centromeres**) to tiny threadlike structures called **spindle fibers** that extend from the opposite sides of the cell.

During migration the chromosomes bend at the centromere, suggesting that they are being dragged by the spindle fibers toward the sides of the cell (see Figure 9.5). Because experiments have shown that muscle fibers can pull objects by contracting, it may be that the spindle fibers also are contracting in order to pull the chromosomes. The statement represents an hypothesis that leads to a prediction as follows:

Hypothesis: *If*...spindle fibers contract in order to pull chromosomes...

Prediction: *then*...the fibers should appear shorter and thicker after the chromosomes have moved.

However, no shortening and thickening of the fibers has been seen; therefore the hypothesis has not been supported.

On the other hand,

Hypothesis: *If*...spindle fibers are necessary for chromosome movement...

and...a chemical known to block the assembly of spindle fibers is injected into dividing cells...

Prediction: *then*...the chromosomes should not move.

When this experiment is performed, the chromosomes, as predicted, do not move; therefore, it appears that the spindle fibers are indeed necessary for chromosome movement.

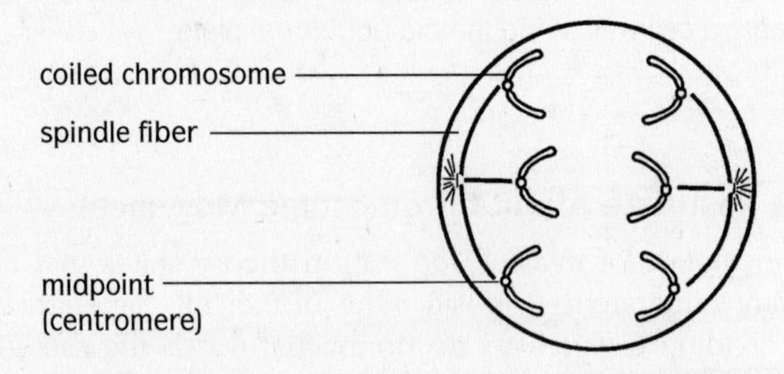

Figure 9.5 During migration, chromosomes bend at the centromere. This occurrence suggests that they appear to be pulled by spindle fibers.

THEORIES OF SEXUAL REPRODUCTION AND MEIOSIS

Most multicellular organisms reproduce sexually; that is, most are either male or female individuals that produce special kinds of cells called **gametes.** The male produces gametes called **sperm** cells and the female produces gametes called **egg** cells. In higher plants, the sperm cells are located in pollen grains. The sperm and egg cells each contain only half of a complete set of chromosomes. When the sperm cell enters the egg cell, its chromosomes combine with those of the egg cell. This is called **fertilization** and the resulting cell with a complete set of chromosomes is called a zygote. The zygote is the first cell of the next generation individual. The new individual develops from this first cell zygote by cell replication (mitosis and cytokinesis) and cell differentiation.

Gamete formation involves a process in which cells that contain one set of chromosomes divide to produce new cells that contain only half of a set of chromosomes. Reduction of the number of chromosomes is necessary to ensure that the zygote will not receive too many chromosomes. This process of gamete formation is called **meiosis** and, like mitosis, is a process that can be thought of as occurring in phases (see Figure 9.6).

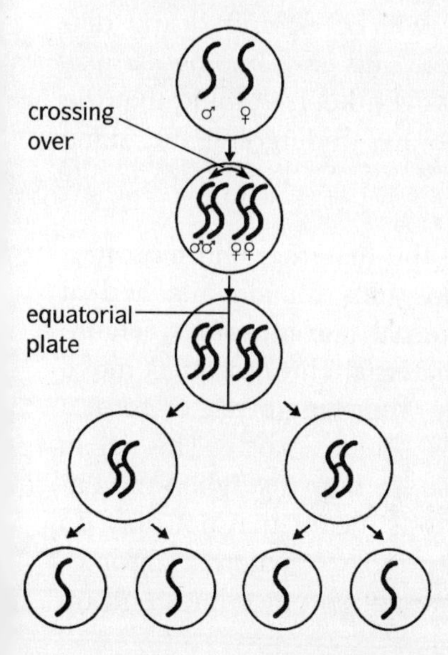

A. Maternal (♀) and paternal (♂) chromosomes exist in like (homologous) pairs (only one pair is shown).

B. Maternal and paternal chromosomes replicate and remain attached at their midpoints. They then move together to form a group of four homologous chromosomes. Some crossing over of maternal and paternal segments may occur at this time.

C. The group of four homologous chromosomes then migrates to the equatorial plate. Two chromosomes end up on each side of the plate.

D. A cell division occurs along the equatorial plate, separating pairs of homologous chromosomes into two new cells.

E. A second cell division occurs, producing four gametes, each with a single chromosome from each of the original group of four homologous chromosomes.

Figure 9.6 The process of meiosis.

Before discussing the process of meiosis in more detail, we need to be clear on just what constitutes a set of chromosomes. In the nucleus of most cells in the body are found groups of chromosomes identical to those found in the other body cells. In sexually reproducing organisms, half of these chromosomes initially came from the mother's egg (the maternal chromosomes) and half came from the father's sperm (the paternal chromosomes). In humans there are 23 maternal chromosomes and 23 paternal chromosomes for a total of 46 chromosomes in the entire set. The total number of chromosomes varies widely from one type of organism to another. For example, an earthworm has a set of 32 chromosomes, a gill fungus has only four chromosomes per set, while an elm tree has 56.

Not all chromosomes within the cell's nucleus, the set, appear the same. Rather, they look like tiny cigars of different lengths, some of which are bent. But, with only one exception, for each paternal chromosome of a particular length and shape, there is a maternal chromosome of the same length and shape. In other words, the paternal and maternal chromosomes of a cell occur in like-pairs. These like-pairs are referred to as **homologous chromosomes.**

The process of meiosis takes place in only those special body cells destined to become the gametes, that is, the egg or sperm cells. For the sake of simplicity, the process of meiosis will be described and diagrammed for just one paternal chromosome and its homologous maternal chromosome (see Figure 9.6A). However, keep in mind that the other homologous pairs of chromosomes are undergoing the same processes at the same time.

Meiosis begins when the paternal and the maternal chromosomes replicate. The replicated paternal chromosomes remain attached at their midpoints, as do the replicated maternal chromosomes (Figure 9.6B). Next, the replicated paternal and maternal chromosomes move together, so that they line up next to each other lengthwise to form a group of four homologous chromosomes (Figure 9.6C). This is an amazing event in and of itself. For example, in humans, when the entire set of 46 chromosomes is considered, this means that not only do the paternal chromosomes end up lined up next to maternal chromosomes, but somehow the paternal and maternal chromosomes end up lined up next to their homologous mates.

During the next phase of meiosis, the group of four homologous chromosomes moves to the equatorial plate of the nucleus (as do other groups of homologous chromosomes not shown in the figure), so that two of the four chromosomes are on one side of the plate and the other two are on the other side. After assembling along the equatorial plate, the two chromosomes on one side of the plate move in one direction, while the two on the other side move in the opposite direction. The cell then divides into two cells with two homologous chromosomes ending up in each cell (Figure 9.6D).

To complete meiosis, the two new cells divide once again in such a way that each of the resulting four cells receives just one chromosome (Figure 9.6E). These four cells are the gametes, that is, the egg or sperm cells, depending upon whether meiosis took place in a female or a male individual.

The Importance of Sexual Reproduction and Meiosis

Sexual reproduction and meiosis result in a mixing of chromosomes of individuals. This mixing produces a greater genetic diversity (variety) of individuals in later generations than would be the case in organisms that do not reproduce sexually and do not produce gametes through meiosis. This greater genetic diversity results in individuals with greater variations in anatomical, physiological, and behavioral characteristics, thus increasing the chances that some members of those subsequent generations will be successful and survive (see Chapter 13).

The mixing of chromosomes takes place three times during sexual reproduction and meiosis. First, when fertilization takes place, chromosomes from the female and from the male parent mix in the zygote, that is, the first cell of the next generation offspring. Second, mixing occurs during meiosis when the paternal and maternal chromosomes have duplicated and remain lined up next to each other in the group of four (Figure 9.6C). At this time, some segments of the duplicated chromosomes may twist around each other, break off, and switch places. This switching of places, technically referred to as **crossing over,** results in some mixing of paternal and maternal chromosomes. The third time mixing occurs is when the duplicated paternal and duplicated maternal chromosomes line up at the cell's equatorial plate. The paternal chromosomes do not all line up on one side of the plate and the maternal chromosomes do not all line up on the other side. Therefore, when the paternal and maternal chromosomes move away from the plate, some of the paternal and some of the maternal chromosomes move together and end up mixed together in each of the resulting cells.

POSTULATES OF SEXUAL REPRODUCTION THEORY

1. Most multicellular organisms are either male or female individuals that produce special kinds of sex cells called gametes (either sperm or eggs).
2. The gametes each contain only half of a complete set of chromosomes.
3. When the sperm cell enters an egg cell, their chromosomes combine to produce a new cell (the zygote) with a complete set of chromosomes.
4. The new individual develops from the zygote through mitotic cell divisions and differentiation.
5. Gametes are produced through a process called meiosis, in which the number of chromosomes are cut in half.
6. Sexual reproduction results in a greater genetic diversity than asexual reproduction, hence a greater chance of the survival of some individuals.

POSTULATES OF MEIOSIS THEORY

1. Meiosis begins when the maternal and paternal chromosomes duplicate.
2. The duplicated maternal and paternal homologous chromosomes then move and line up next to each other. This results in a group of four homologous chromosomes.
3. At this time, some chromosome segments may cross over each other and switch positions.
4. The groups of four homologous chromosomes migrate to the equatorial plate, so that two chromosomes of each group are on each side of the plate.
5. Two of the homologous chromosomes, of each group of four, now move away from the plate in one direction, while the other two, of each group, move in the opposite direction.
6. A new cell membrane forms to produce two new cells with two homologous chromosomes, of each group of four, in each new cell.
7. Each of these two new cells undergoes another cell division, so that each of the four resulting cells, the gametes, receives just one chromosome from each of the original groups of four homologous chromosomes (see postulate 2).

The Link Between Mendel's "Factors" and Meiosis

During the 1870s and 1880s a number of biologists worked out accurate descriptions of the process of meiosis. At that time, the relationship between Mendel's theory of inheritance and meiosis was not known. However, in 1902 and 1903, this changed when Walter Sutton, a young graduate student working at Columbia University in New York, proposed an hypothesis that linked the two. Sutton's hypothesis was that Mendel's "factors" (that is, genes), which presumably carried the hereditary instructions from parents to offspring, are located in the chromosomes in the egg and sperm cells of the parents.

Sutton did more than propose the hypothesis. He provided an elegant argument in its favor. Sutton's argument can be summarized in four basic steps:

Step 1

Hypothesis: *If*...genes are carried by (i.e., located in) chromosomes...

and...as Mendel's theory states, egg and sperm cells have half the number of genes that body cells have...

Prediction 1: *then*...egg and sperm cells should have half the number of chromosomes that body cells have.

Step 2

Hypothesis: *If*...genes are carried by chromosomes...

and...as Mendel's theory states, gene pairs separate during egg and sperm cell formation...

Prediction 2: *then*...chromosome pairs should separate during egg and sperm cell formation.

Step 3

Hypothesis: *If*...genes are carried by chromosomes...

and...as Mendel's theory states, in fertilization egg and sperm cells unite, restoring the original number of genes...

Prediction 3: *then*...in fertilization egg and sperm cells should unite, restoring the original number of chromosomes.

Step 4

Hypothesis: *If*...genes are carried by chromosomes...

and...as Mendel's theory states, the individual genes are unchanged from one generation to the next...

Prediction 4: *then*...individual chromosomes should remain unchanged from one generation to the next.

Observations of chromosomes made at the time and since Sutton proposed his hypothesis have confirmed Sutton's predictions; therefore, his hypothesis has been supported. Today scientists believe that genes are, indeed, contained in chromosomes.

TESTING HYPOTHESES ABOUT EMBRYOLOGICAL DEVELOPMENT

How does a single cell develop into the millions of highly differentiated cells present in the newborn? There are at least two basic ways that embryological development might occur. First, the egg or the sperm cells could contain a tiny, already-formed individual that simply grows larger with time. This possibility is referred to as the **preformation hypothesis.** Second, the initial cells might start out generally all the same and then duplicate and change to become different specialized types to form various structures during development. This possibility is referred to as the **epigenesis hypothesis.**

Which hypothesis is correct? Probably the first observer of embryological development was Aristotle. Aristotle carefully observed the development of chicken eggs and concluded that epigenesis was correct. It was not until the sixteenth century that people again began to think about and observe embryological development. These people were influenced by passages in the Bible that referred to the "seeds" of man, which suggested to them that, like the tiny preformed leaves in plant seeds, the human sperm and an egg cell also contain tiny preformed individuals. Thus, they generally agreed with the preformation hypothesis and looked for evidence to support it.

Evidence to support the hypothesis of preformation was not hard to find. For example, to some people, chicken eggs seemed to show tiny preformed chickens inside. Aphid eggs were found to develop without being fertilized, suggesting that the eggs must contain preformed aphids. Also, as mentioned, plant seeds were found to contain a tiny embryo that appeared to consist of a tiny root, stem, and leaves. One seventeenth-century observer even imagined that he saw a tiny preformed person curled up in the head of a sperm cell.

Of course, not all scientists at the time agreed that these observations must be explained by resorting to preformation. For example, it was argued that if there was a preformed individual in the sperm, then there must also be another preformed individual inside that individual, and another in that one, and so on. Belief in preformation thus demanded a belief in a series of progressively smaller and smaller individuals, each one contained in the other. To some, this was simply too improbable to accept.

Putting the Postulates to the Test Experimentally

By the nineteenth century, it became clear that mere observation of developing embryos would not allow a clear test of the alternatives. A clear test would require experimentation.

One of the first experiments was conducted by Hans Driesch (1867–1941). Driesch, using just fertilized sea urchin eggs, waited until the zygote divided into two cells. He then completely removed one of the cells and carefully observed what happened to the development of the remaining cell.

The reasoning that guided his experiment was as follows:

Hypothesis: *If*...the zygote contains a tiny preformed sea urchin...

and...one of the first two cells in development is removed...

Prediction: *then*...the remaining cell should develop into only half of a sea urchin.

On the other hand,

Hypothesis: *If*...the zygote and the other cells in the early part of development are undifferentiated...

and...one of the first two cells in development is removed...

Prediction: *then*...the remaining cell should develop into a complete sea urchin.

In other words, the preformation hypothesis leads to the prediction that only half of a sea urchin will develop. Since the remaining cell grew into a complete sea urchin, Driesch concluded that the hypothesis of preformation was wrong and epigenesis was correct.

However, before you hastily agree with Driesch, you should consider the results of a similar experiment by Wilhelm Roux (1850–1924). Working with frog eggs, Roux, like Driesch, allowed the fertilized egg to divide to form two cells. Then using a needle, he killed one of the cells and observed the development of the remaining cell. Just as predicted by the preformation hypothesis, the remaining egg developed into only half of a normal embryo!

Does this mean that preformation is correct after all? Actually, Roux did not come to this conclusion. Rather, he concluded that the result supported the idea that, although tiny, preformed individuals do not really exist, the developing cells quickly differentiate into ones that will become specific parts of the adults when development is complete. Most likely the difference between his results and those of Driesch can be explained by assuming that cell differentiation takes place sooner in frog embryos than it does in sea urchin embryos.

Today biologists think that the basic idea of preformation is false. Powerful electron microscopes have allowed people to see objects as small as individual atoms, yet no preformed individuals have been found. However, a modified version of the preformation hypothesis remains. In a sense, all the information necessary for complete development of the embryo is contained in the set of DNA molecules present in that first fertilized egg cell!

How can one set of instructions present in one cell guide the development of the multitude of different kinds of cells that eventually arise? The modern explanation involves the following key ideas: As new cells arise by mitotic divisions, their environment changes (i.e., an initial cell may be surrounded by water while a later cell may be surrounded by other cells). Environmental changes cause changes in the materials available to the cell as well as the cell's ability to transport materials away from itself. These changes act as signals to the developing cells, which activate or deactivate different parts of their genetic instructions. The differently activated segments of the genetic instructions then direct the development of different kinds of cells.

A MODERN THEORY OF EARLY EMBRYOLOGICAL DEVELOPMENT

Although early development varies somewhat from one species to another, a modern-day theory of embryological development of complex multicellular organisms generally has many commonalities.

POSTULATES OF THE THEORY OF EARLY EMBRYOLOGICAL DEVELOPMENT

1. Embryological development normally begins with fertilization of an egg cell by a sperm cell in which the genetic material of both combine in a single cell called the zygote.
2. Following fertilization, the zygote undergoes a mitotic division to produce two new cells.
3. Mitotic divisions continue, producing four cells, then eight, then sixteen and so on, eventually resulting in a hollow ball of many cells called a **blastula.**
4. An infolding of cells starts to form on the surface of the blastula. The infolding, or *invagination,* of cells continues until the invaginated layer of cells lies against the inside of the outer layer, forming a two-layer ball of cells called the **gastrula.** The outer layer of cell is called the **ectoderm,** and the inner layer is called the **endoderm.**
5. Depending upon the species, either the endoderm cells or the ectoderm cells divide to form a new layer of cells called the **mesoderm,** which lies between the ectoderm and endoderm.
6. The cells, which to this point are very much alike in appearance and all of which contain a full set of the genetic instructions that were present in the zygote, begin to differentiate and specialize and to take on different structures and functions, depending upon their locations (i.e., environment) in the embryo. (For example, ectoderm becomes the skin, mesoderm becomes the internal organs, endoderm becomes the digestive tract.) Cell differentiation and specialization occur because the part of the genetic instructions that are active in any particular cell vary, depending upon its environment and the materials available to it.
7. Embryological development is complete when the new individual is born, yet cellular differentiation and growth continue to take place in most species.

QUESTIONS

1. Notice that Hammerling's experiments were conducted on a single type of cell. Notice also that the general conclusion drawn was that the cell nucleus controls virtually all cell activities. What

problem(s) do you see in drawing a general conclusion about all cells from experiments on just one type of cell? What can/should be done to reduce these problems?

2. Propose an hypothesis to explain how spindle fibers cause chromosomes to move.

3. How might your hypothesis be tested?

4. All of the available evidence supports Sutton's hypothesis that genes are located on chromosomes. Can we, therefore, conclude that Sutton's hypothesis has been proven? Explain.

5. Suppose sexually reproducing organisms did not produce gametes by meiosis. In other words, suppose no reduction of the number of chromosomes occurred prior to fertilization. This would result in a zygote with twice the number of chromosomes

as either of its parents. Why might this either immediately or after a few generations become a problem for the organism?

6. What is wrong with the statement that the development of any particular characteristic (i.e., the color of a flower) is controlled by the genes?

7. What is wrong with the statement that the development of any particular characteristic is controlled by the organism's environment?

TERMS TO KNOW

blastula	gamete
cell replication	gastrula
centromere	homologous chromosome
crossing over	meiosis
cytokinesis	mesoderm
ectoderm	mitosis
egg	nucleus
endoderm	preformation hypothesis
epigenesis	sperm/pollen
fertilization	spindle fibers

SAMPLE EXAM

Circle or enter the letter of the best answer (e.g., a, b, c, d). For items 1–5, use cell shape to match the animal cells in Figure 9.7 labeled A–E with their functions.

1. Cell travels around the body in tubes to transport molecules that enter and exit the cell through its cell membrane.

____

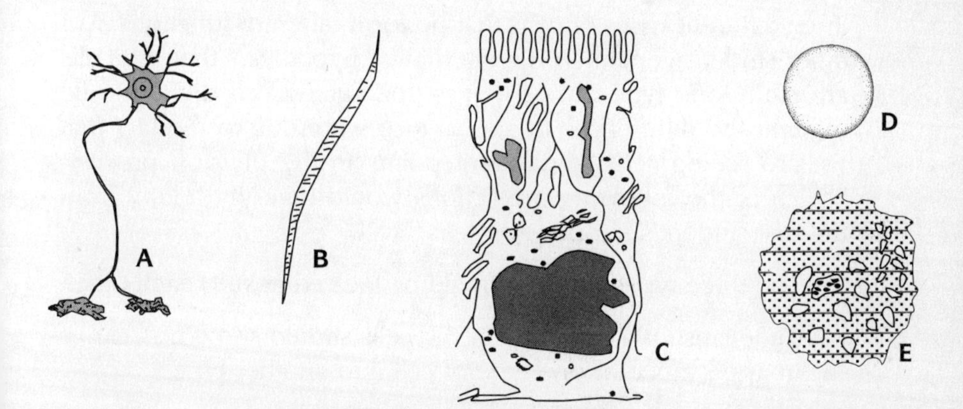

Figure 9.7 A variety of animal cells.

2. Cell picks up electrical signals from a number of other cells and transmits a common signal to another cell or cells.

____

3. Cell is capable of constructing lengthwise to move such structures as bones.

____

4. Cell is found in skin and can expand and contract to change skin color.

____

5. Cell is found on the inside of the intestinal wall and absorbs food molecules from the digestive tract.

____

6. The cells shown in Figure 9.8 were observed on an onion root tip slide.

Figure 9.8 Cells observed on an onion root tip slide.

One student generated the hypothesis that the cells are really three different types of cells that perform different functions. Another student generated the alternative hypothesis that the cells are really one type of cell that is undergoing cell division. She thought the differences in appearance were due to the fact that we are seeing cells at different points in the division process. Which of the following predictions would logically follow from the first student's hypothesis?

a. The three types of cells should be located next to each other.

b. Time-lapse photography of the cells should show no change in appearance of one type of cell into another type.

c. The three types of cells should not be located next to one another.

d. Time-lapse photography of the cells should show a change in appearance of one type of cell into another type.

e. The size of the cells should not vary over time.

7. Which of the predictions mentioned in number 6 would follow logically from the second student's hypothesis?

a. The three types of cells should be located next to one another.

b. Time-lapse photography of the cells should show no change in appearance of one type of cell into another type.

c. The three types of cells should not be located next to one other.

d. Time-lapse photography of the cells should show a change in appearance of one type of cell into another type.

e. The size of the cells should not vary over time.

8. The significance of the greater amount of yolk in birds' eggs as compared to the amount of yolk in mammalian eggs is that

a. birds need more energy for development.

b. mammals do not depend totally on the yolk for development.

c. birds develop more slowly and therefore need a greater supply of yolk.

d. the yolk of mammals is more concentrated.

9. To test the hypothesis that flower petals produce a chemical necessary for fruit development, a botanist removed all the petals from the flowers on one plant. She left the petals intact on a similar plant. What prediction logically follows from her hypothesis and experimental design?

a. Fruits will develop regardless of whether the petals have been removed.

b. If petals produce a chemical, then the chemical is necessary for fruit development.

c. What causes fruit to develop?

d. Fruit should develop only from the flowers with petals.

e. Fruit develops from the flower's ovary, not from its petals.

10. Suppose in number 9 the botanist found that on the plant with petals removed, 6 of its 20 flowers developed into fruit, while on the plant with its petals intact, 10 of its 30 flowers did. What conclusion can you reasonably draw?

    a. The petal hypothesis has been supported, because 4 more flowers were produced on the plant with petals intact.

    b. Fewer than half of the flowers developed into fruits in both experimental conditions.

    c. The experiment should be repeated with better controls before a conclusion can be reached.

    d. Petals produce a chemical necessary for fruit development.

    e. The petal hypothesis has not been supported, because approximately the same percentage of fruits developed in both conditions.

For questions 11 and 12, assume that Figure 9.9 represents the eyepiece and objectives of a microscope. Under lower power magnification, 100 yeast can be counted side by side along the diameter of the field.

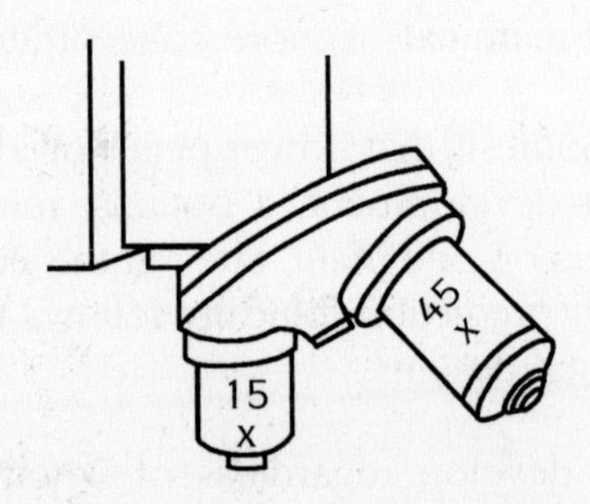

Figure 9.9 Microscope objectives.

11. About how many of these same yeast cells will you see when you turn the nosepiece to high power?

    a. 3

    b. 33

    c. 130

    d. 145

    e. 300

12. With a 10X eyepiece, the yeast under low power will be magnified

    a. 10X.

    b. 55X.

    c. 150X.

    d. 450X.

    e. 675X.

CHAPTER 10

GENERAL MOLECULAR LEVEL THEORIES

GETTING FOCUSED

- *Learn about kinetic molecular theory.*
- *Discover how the hypothesis that atoms exist was tested.*
- *Learn about general metabolic theory.*
- *Evaluate a theory of molecular movement.*

KINETIC-MOLECULAR THEORY

The universe consists of matter, empty space, and light. Pieces of matter move about in this empty space and occasionally collide, which can cause them to change their speed and/or direction of motion. Matter in motion is said to have **energy,** the energy of motion. When one piece of matter hits another piece of matter, its energy of motion (**kinetic energy**) can be transferred to that other piece. For example, when one pool ball squarely strikes another, the first ball stops and the second ball moves. The total amount of kinetic energy (motion) is not reduced, it is merely transferred from one ball to the other.

All of the matter in the universe consists of combinations of tiny particles called atoms. Atoms consist of subatomic (inside the atom) particles called protons, neutrons, and electrons. The protons and neutrons of atoms are found in the center (the atomic nucleus), while the much smaller electrons move rapidly around the nucleus. A force exists that attracts the electrons to the protons in the nucleus. To explain this force, protons are assumed to carry a positive charge while electrons are assumed to carry a negative charge, and opposite charges are assumed to attract. Although the actual cause of the force is not known, the attractive force is normally opposed by a force of the electron's motion away from the nucleus (centripetal force), so that the electrons normally remain at a specific distance from the nucleus. But these opposing forces are not always equal, so electrons may move closer or farther from the nucleus, or they may move from one atom to another.

Although the precise nature of light is not known, at the present let's simply say that light appears to consist of tiny particles. These tiny particles are called *photons*. Photons travel through space very rapidly. Photons that interact with electrons may cause the electrons to move more rapidly. Photons may also be released from the electrons. When this happens, light is given off and the electrons lose some of their motion.

The Atomic Level

Ninety-two kinds of atoms have been found to occur naturally on Earth. They differ from one another in their numbers of protons, neutrons, and electrons. The smallest and lightest kind of atom (hydrogen) has one proton and one electron. Hydrogen atoms tend to combine in groups of two, so that two electrons move around two atom nuclei. The next heaviest atom, helium, has two protons, two neutrons, and two electrons. Lithium has three protons, three neutrons, and three electrons.

The pattern of increasing numbers of protons, neutrons, and electrons continues through the remaining kinds of atoms. But in some of the larger and heavier atoms, the number of neutrons is not always equal to the number of protons or electrons. Some familiar atoms are carbon with 6 protons, 6 neutrons, and 6 electrons; iron with 26 protons, 26 electrons and, normally, 28 neutrons; and gold with 79 protons, 79 electrons and, normally, 118 neutrons. The heaviest of the naturally occurring atoms, uranium, is composed of 92 protons, 92 electrons and, normally, about 146 neutrons. Even though uranium atoms are the heaviest naturally occurring atoms, and are moderately sized, it would still take about 76 million of them placed side by side to span the diameter of a penny. Particular kinds of atoms in which the number of neutrons may vary are called *isotopes.*

When groups of atoms are very close and do not move past one another very easily due to forces that hold them in place, they are called *solids.* When the atoms are not packed so tightly together and can move past one another with relative ease, but are still attached to one another, they make up substances known as *liquids.* Mercury is a type of liquid made up of just one kind of atom. When the atoms of a structure are so far apart that they move past one another very rapidly and occasionally collide and bounce in various directions, the substance is referred to as a *gas.* At room temperature, hydrogen and helium are gases.

The Molecular Level and Chemical Reactions

Most of the 92 different kinds of atoms that naturally occur in nature exist in various combinations with other atoms. Atoms combine with one another because of forces of attraction called chemical bonds. Combinations of atoms, formed by chemical bonds among the atoms, are called *molecules.* Molecules can be very simple such as two hydrogen atoms that are bonded, or very complex, such as the hundreds of atoms bonded in complex molecules found in living things, such as protein, carbohydrate, lipid, and nucleic acid molecules.

The chemical bonds of molecules keep the atoms of molecules from flying apart. That is, they store up the potential motion of the atoms. When bonds are struck by rapidly moving objects such as photons, electrons, atoms, or molecules, they may "break" and the atoms may fly away from each other, somewhat like the way a rack of pool balls move apart when the cue ball hits them. The stored motion (i.e., **potential energy**) is "released" and the moving atoms can move about and strike other atoms or molecules, perhaps causing them to break apart or perhaps themselves combining with another atom or molecule. Interactions in which the bonds holding the atoms of a molecule together are broken or new bonds are formed to create new molecules are collectively referred to as **chemical reactions.** For example, a chemical reaction takes place when an electrical current (flow of electrons) passes through liquid water (H_2O) and breaks the bonds between the individual hydrogen (H) and oxygen (O) atoms. New bonds between individual hydrogen atoms and between individual oxygen atoms are then formed. The result is that hydrogen molecules (H_2) and oxygen molecules (O_2) are given off as gases. On the other hand, when sugar molecules are dissolved in water, the bonds among the atoms in the individual sugar ($C_6H_{12}O_6$) and water molecules (H_2O) do not break apart. The sugar and water molecules merely mix with each other. Therefore, dissolving sugar molecules in water in not a chemical reaction.

A flame gives evidence of a very rapid chemical reaction. In a flame, molecular bonds are breaking rapidly and atoms and/or molecules are rapidly flying apart. Photons are also being released. The **temperature** of a substance (solid, liquid, or gas) is a measure of the amount of motion of its atoms and/or molecules. The more rapid the motion, the higher the temperature.

The term pressure refers to the collective force exerted on a surface by atoms and/or molecules that strike that surface. More particles, larger particles, and higher speeds cause greater pressure.

POSTULATES OF THE KINETIC-MOLECULAR THEORY OF MATTER

1. Matter consists of small particles (atoms and combinations of bonded atoms called molecules) and light, which consists of still smaller particles called photons.
2. Matter moves and can strike other matter and transfer some/all of its motion (kinetic energy) to the other piece of matter.
3. Photons can interact with electrons and cause them to move more rapidly. Photons may also be "released" from electrons, which causes light to be emitted and results in a reduction in the motion of the electrons.
4. Atoms differ from one another in the number of protons, neutrons, and electrons each one contains.
5. Attractive forces between atoms (molecular bonds) can be broken, causing the atoms to move apart, which in turn can cause collisions and transfers of energy.
6. Molecular bonds can be formed between atoms when they strike one another.
7. An energy source, such as a flame, consists of rapidly moving particles (atoms/molecules, photons) that can transfer some of their motion energy to nearby particles through collisions.
8. The temperature of a substance is a measure of the amount of motion of its particles (that is, the more motion the greater the temperature).
9. The term pressure refers to the force exerted on a surface by the collisions of particles (that is, more and larger particles at higher speeds equals greater air pressure).

Testing the Hypothesis That Atoms Exist

In 450 B.C. Democritus, a Greek philosopher, hypothesized that matter is composed of tiny invisible and indivisible particles. He named these hypothetical particles *atoms,* from the Greek word *atomos,* which means "indivisible." Although Democritus's idea seemed like a good one to many people through the years, it was not until the work of John Dalton (1776–1844), an English school teacher, that convincing evidence was obtained to support the hypothesis that atoms exist. Let's take a brief look at some of Dalton's reasoning and evidence.

Dalton hypothesized that substances were composed of indivisible particles that weighed different amounts. He also hypothesized that they could combine in different ways. For example, one particle (atom) of one substance (A) could combine with one atom of another substance (B) to form a compound like this: (A)(B). He imagined other ways they could combine, such as these (A)(B)(B), (A)(B)(A), and so forth.

The important point in terms of the theory of atoms is this: If matter is really composed of atoms, then these composites should always consist of whole-number ratios, because atoms were supposed to be indivisible. In other words, if fractions of atoms do not exist, then fractions of atoms in combinations with others also do not exist. For example, one atom of A could combine with one atom of B, but atom A could not combine with, say, half of atom B.

Dalton's hypotheses about atoms, therefore, lead one to predict that the weights of various combinations of two substances should be in fairly simple whole-number ratios. The only problem was that individual atoms could not be weighed. Nevertheless, let us look at some of Dalton's data to see if they fit his theory.

In 1810, Dalton published the data on five gases that were thought to be compounds (combinations) of atoms of nitrogen (N) and oxygen (O). Table 10.1 lists the data. Note that the name of the gases in quotations are not the names used today.

TABLE 10.1 DALTON'S FIVE GASES

		PERCENTAGE BY WEIGHT	
COMPOUND	**RELATIVE DENSITY**	**NITROGEN**	**OXYGEN**
"Nitrous gas" (NO)	12	42	58
Nitrous oxide	17	59	41
"Nitric acid"	19	27	73
"Oxynitric acid"	26	20	80
"Nitrous acid"	31	33	67

Because nitrous gas was the lightest compound, Dalton assumed that it consisted of molecules of one nitrogen atom combined with one atom of oxygen. The percentages by weight that he obtained when he decomposed nitrous gas into nitrogen and oxygen were 42 for nitrogen and 58 for oxygen. Since he had assumed that nitrous gas contained an equal number of nitrogen and oxygen atoms, he concluded from this that one nitrogen atom weighed about 4 units to about 6 units for one oxygen atom (i.e., 42 units to 58 units is about 40 units to 60 units or 4 units to 6 units).

Dalton then guessed that the next gas, nitrous oxide, consisted either of two atoms of nitrogen for every one atom of oxygen or one atom of nitrogen for every two atoms of oxygen. The hypothesis of two atoms of nitrogen (4 + 4 = 8 units of weight) and one atom of oxygen (6 units of weight) led Dalton to predict that the ratio of weights of nitrogen to oxygen in the decomposed gas would be 8 nitrogen to 6 oxygen or 8/14 = 57% nitrogen to 6/14 = 43% oxygen. The hypothesis of one atom of nitrogen (4 units) and two atoms of oxygen (6 + 6 = 12 units) led Dalton to predict that the ratio of weights of nitrogen to oxygen in the decomposed gas would be 4 nitrogen to 12 oxygen or 4/16 = 25% nitrogen to 12/16 = 75% oxygen. The actual percentages in the table are 59% for nitrogen and 41% for oxygen. Because these percentages are much closer to those predicted by the hypothesis of two atoms of nitrogen to one atom of oxygen, that hypothesis is supported. Also it should be pointed out that the result also supports the more general postulate that matter is composed of distinct indivisible particles (i.e., atoms) because, as this hypothesis predicts, his results and analysis indicate that atoms combine in simple, whole-number ratios.

Notice that the percentage by weight of nitrogen and oxygen for Dalton's third gas, nitric acid, are 27% nitrogen to 73% oxygen. Because these percentages are so close to those predicted by the hypothesis discussed above, one atom of nitrogen and two atoms of oxygen, it seems likely that nitric acid is a gas made up of molecules that contain one atom of nitrogen bonded to two atoms of oxygen.

Based upon this sort of reasoning, see if you can now generate hypotheses about the nature of the molecules that make up the remaining two gases studied by Dalton, oxynitric acid and nitrous acid. Do the percentages shown in Table 10.1 support or contradict your hypothesized structures? What is the most likely molecular structure of these two gases? Explain.

____

____

____

____

Why then do scientists believe that atoms exist? The answer is that the hypothesis that atoms exist leads to predicted results that have, in fact, been found. The reasoning and evidence can be summarized as follows:

Hypothesis: *If*...matter consists of indivisible particles that have specific weights and combine with one another in specific ways...

and...combinations of atoms are decomposed into their parts...

Prediction: *then*...the ratios of weights of those parts should be in simple, whole-number ratios.

Result: The ratios of weights of those parts are in simple, whole-number ratios.

Conclusion: *Therefore*...the hypothesis has been supported.

GENERAL METABOLIC THEORY

Kinds of Atoms and Molecules in Living Things

Although nearly one hundred different kinds of atoms have been found naturally occurring on Earth, only six make up some 99% of all living things. These six are carbon (C), hydrogen (H), nitrogen (N), oxygen (O), phosphorous (P), and sulfur (S)—easily remembered as CHNOPS.

CHNOPS combine in various ways to make up the molecules found in living things. The most common combination of atoms is two hydrogen atoms bonded to one oxygen to form water (H_2O). Water makes up 50–95% of cells by weight. In addition to water, living things consist primarily of four major classes of molecules called carbohydrates, lipids, proteins, and nucleic acids. These molecules are sometimes referred to as *organic* molecules because they are constructed by and occur in organisms. Molecules that occur in nonliving systems are sometimes referred to as *inorganic* molecules.

Carbohydrates are a class of organic molecules composed of combinations of hydrogen, oxygen, and carbon atoms. In most carbohydrates, there are two hydrogen atoms and one oxygen atom for every one carbon atom. Sugars, starches, and cellulose are types of carbohydrates. A diagram of a common type of sugar called glucose ($C_6H_{12}O_6$) is shown in Figure 10.1. Starches consist of simpler sugars, like glucose, which are bonded.

Lipids, like carbohydrates, are composed principally of hydrogen, oxygen, and carbon atoms. But lipids differ from carbohydrates in that they contain a much smaller proportion of oxygen atoms, and they may also contain atoms of phosphorous and nitrogen. Fats, oils, waxes, and steroids, such as cholesterol, are all lipids.

Proteins are a third class of molecule that consist of combinations of hydrogen, oxygen, carbon, and nitrogen atoms. Often they also contain sulphur atoms. Protein molecules are very complex, but all consist of

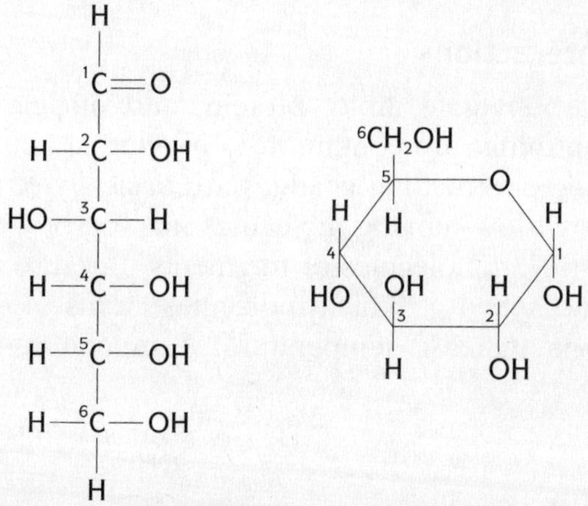

Figure 10.1 Two forms of glucose. Glucose molecules may exist in a straight chain form as shown on the left, and more commonly as a ring form as shown on the right.

chains of two or more basic units called **amino acids.** There are 20 different kinds of amino acids, all have the basic form shown in Figure 10.2.

The R represents a side branch of one or more atoms that differ from one amino acid to the next. For example, the simplest amino acid, glycine, has a side branch of a single hydrogen atom. The amino acid called alanine has a side branch of one carbon and three hydrogen atoms. Other amino acids have progressively more complex side branches.

Protein molecules such as collagen, insulin, and hemoglobin consist of long chains of amino acids in different lengths, shapes, and sizes. For example, the chain of amino acids in insulin is shown in Figure 10.3, while Figure 10.4 shows how the four separate chains of amino acids in a molecule of hemoglobin are folded around one another. Protein molecules that are folded like hemoglobin are called globular proteins. Enzymes (see below) are globular proteins.

Nucleic acids are the fourth major class of molecules found in living things. They are the molecules of which genes, the units of heredity, are composed. The messenger molecules that convey information from the genes to the rest of the cell are also nucleic acids. Like complex carbohydrates and proteins, nucleotides are complex molecules that consist of chains of smaller building blocks. The building blocks of nucleic acids are called **nucleotides.** Nucleotides consist of still smaller parts: a five-carbon sugar; a small group of atoms with a phosphorous atom at the center, called a phosphate group; and a small group of atoms containing nitrogen.

Molecular Interactions

The water, carbohydrate, lipid, protein, and nucleic acid molecules found in living things are continuously moving about and undergoing structural changes. Structural changes are basically of two kinds. First, larger, complex, slow-moving molecules may be broken apart to form smaller, fast-moving molecules or fragments. Because temperature is a measure of the speed at which molecules/atoms move, these breakdown reactions increase temperature. Temperature-increasing reac-

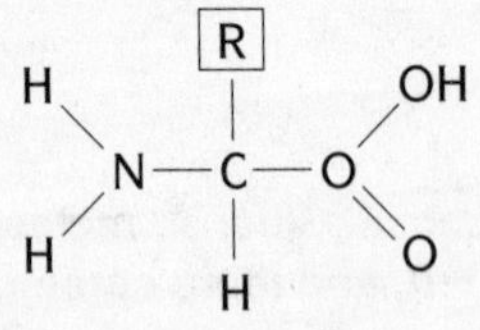

Figure 10.2 The basic structure of an amino acid.

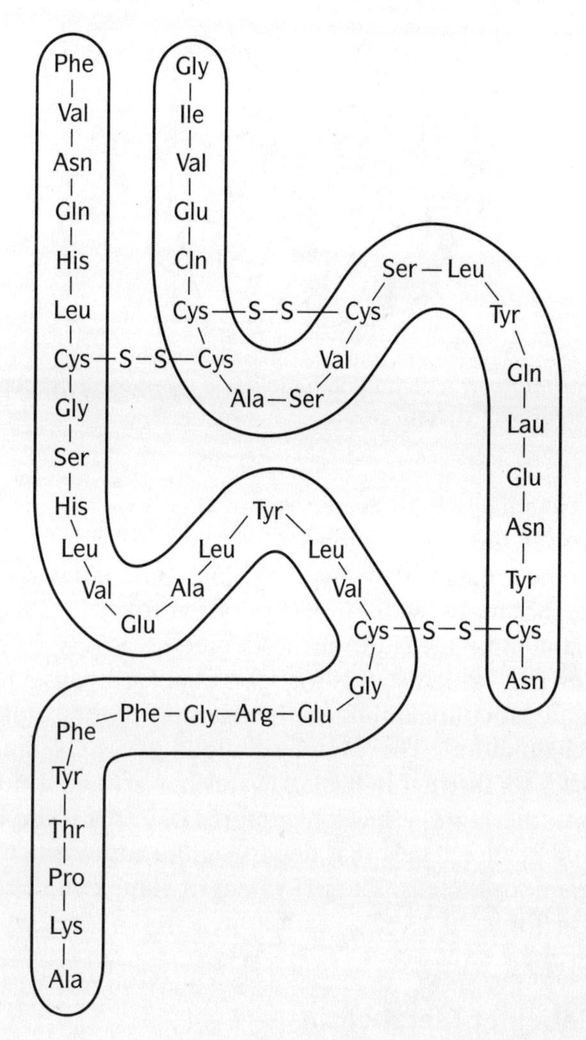

Figure 10.3 The structure of insulin. The insulin molecule consists of two chains of linked amino acids joined by two bonds made of two sulfur atoms (shown as an S-S). There is also one disulfide bond within the shorter chain (right).

tions are called **exothermic** reactions. Second, fast moving molecules and/or fragments may combine to form larger, more complex and slower-moving molecules. These temperature-decreasing reactions are called **endothermic** reactions.

The chemical reactions involved in breaking apart or building molecules are collectively referred to as the organism's **metabolism.** The rate of these chemical reactions is referred to as **metabolic rate.** A very common type of molecular breakdown may occur when complex

Figure 10.4 Structure of a hemoglobin molecule composed of four linked amino acid chains, each with several twists.

molecules are mixed with water. Many break apart (dissociate) into smaller parts. If the dissociation of a molecule adds more hydrogen ions (H^+) to the solution than hydroxyl ions (OH^-), the solution is said to be *acidic*. Solutions with an excess of hydroxyl ions are said to be *basic*. A quantitative measure of the degree of acidity or basicity of a solution is its **pH,** which stands for "potential for hydrogen" and is defined as the negative logarithm, to the base 10, of the solution's hydrogen ion concentration. The pH of solutions generally ranges from O (highly acidic) to 14 (highly basic). Water has an equal number of H^+ and OH^- ions; therefore, it has a neutral pH of 7. pH is important to living systems, because levels that depart significantly from 7 disrupt the chemical reactions of cells. The pH values of some common substances are shown in Table 10.2.

The Regulation of Metabolism

A mixture of hydrogen and oxygen atoms does not react; but if a spark is provided, the mixture will react in an explosion! An explosion of hydrogen and oxygen atoms will also occur if a small piece of platinum is added. After the explosive reaction is over, the platinum will still be present, unchanged.

A substance, such as platinum, that speeds up a chemical reaction but is itself unchanged when the reaction is over (even though it may have been temporarily altered during the reaction) is known as a **catalyst.** A catalyst affects only the rate of reaction. It simply speeds up a reaction that is initially possible. Catalysts decrease the energy/motion needed for a reaction to take place. This needed energy is called the **activation energy** of the reaction. The catalyst reduces the activation

TABLE 10.2 APPROXIMATE pH OF SOME COMMON SUBSTANCES

Substance	pH
Apples	2.9–3.3
Apricots (dried)	3.6–4.0
Asparagus	5.4–5.7
Beans	5.0–6.0
Beer	4.0–5.0
Beets	4.9–5.6
Blackberries	3.2–3.6
Bread, white	5.0–6.0
Cabbage	5.2–5.4
Carrots	4.9–5.2
Cherries	3.2–4.1
Cider	2.9–3.3
Corn	6.0–6.5
Crackers	7.0–8.5
Dates	6.2–6.4
Flour, wheat	6.0–6.5
Ginger ale	2.0–4.0
Gooseberries	2.8–3.1
Grapefruit	3.0–3.3
Grapes	3.5–4.5
Hominy	6.9–7.9
Human blood plasma	7.3–7.5
Human duodenal contents	4.8–8.2
Human feces	4.6–8.4
Human gastric contents	1.0–3.0
Human milk	6.6–7.6
Human saliva	6.0–7.6
Human spinal fluid	7.3–7.5
Human urine	4.8–8.4
Jams, fruit	3.5–4.0
Jellies, fruit	3.0–3.5
Lemons	2.2–2.4
Limes	1.8–2.0
Magnesia, milk of	10.5
Milk, cow	6.4–6.8
Molasses	5.0–5.4
Olives	3.6–3.8
Oranges	3.0–4.0
Peaches	3.4–3.6
Pears	3.6–4.0
Peas	5.8–6.4
Pickles, dill	3.2–3.5
Pickles, sour	3.0–3.5
Pimento	4.7–5.2
Plums	2.8–3.0
Pumpkins	4.8–5.2
Raspberries	3.2–3.7
Rhubarb	3.1–3.2
Salmon	6.1–6.3
Sauerkraut	3.4–3.6
Shrimp	6.8–7.0
Spinach	5.1–5.7
Squash	5.0–5.3
Strawberries	3.1–3.5
Sweet potatoes	5.3–5.6
Tomatoes	4.1–4.4
Tuna	5.9–6.1
Turnips	5.2–5.5
Vinegar	2.4–3.4
Water, distilled (carbon dioxide free)	7.0
Water, mineral	6.2–9.4
Water, sea	8.0–8.4
Wines	2.8–3.8

energy needed by orienting the reactants to each other in such a way that important internal bonds are weakened, thus making them easier to break (e.g., by being struck by other molecules).

A catalyst, such as platinum, is rather unselective about the reactions it speeds up. Instead of using unselective substances such as platinum, living things contain a huge variety of large globular proteins that act as catalysts. These organic catalysts are called **enzymes.** There are many kinds of enzymes. Each regulates a specific chemical metabolic reaction. For example, an enzyme called salivary amylase, produced in the salivary glands, speeds up the breakdown of complex starch molecules into smaller glucose molecules so that they are small enough to be transported into the blood stream and enter the body's cells.

POSTULATES OF THE GENERAL THEORY OF METABOLISM

1. Living things consist primarily of six kinds of atoms (CHNOPS) that combine in various ways to produce water and four classes of molecules called carbohydrates, lipids, proteins, and nucleic acids.
2. The molecules found in living things continuously move and undergo structural changes of two basic types: (a) large, slow-moving molecules break apart to form small, fast-moving molecules (exothermic reactions), and (b) small, fast-moving molecules and/or fragments combine to form larger, slower-moving molecules (endothermic).
3. The rate of these chemical reactions is partially regulated by protein molecules called enzymes.
4. Enzymes increase the rate of a chemical reaction but are not structurally changed themselves; therefore, they can be used repeatedly.
5. Each chemical reaction is governed by an enzyme specific to that reaction.

A THEORY OF MOLECULAR MOVEMENT

When sunlight shines on such objects as your skin, they get hotter; that is, their temperature increases. Temperature increases because the atoms and/or molecules of the object begin to move more rapidly. If the molecules in your skin start moving too fast, they can be broken apart. Your skin literally burns. This, of course, is called a sunburn.

Not only can the motion of the molecules in your skin be traced to the transfer of motion from incoming photons of light, nearly all the motion on Earth can be directly or indirectly traced to collisions with these photons. If the input of sunlight were cut off, forces of attraction among molecules would cause them to cluster, thus all of this motion would eventually stop. The temperature would drop to absolute zero. But the input of sunlight has not been cut off. The sunlight causes objects to move, which causes other objects to move, and so on. What sorts of factors are important to the motion of the molecules in living things?

Diffusion

Imagine a small box containing 25 marbles clustered at one end. When the box is shaken, the marbles collide with one another and the walls of the box. Clearly, of all the movements a single marble may take, more lead away from the center of the cluster than toward it; therefore, random movement of the marbles resulting from shaking of the box will tend to disrupt the cluster rather than maintain it. Eventually, the moving marbles will be dispersed over the entire bottom of the box.

At the molecular level, such movements of particles from areas of high concentration to areas of low concentration resulting from being struck by photons of light and/or their random collisions with other molecules is called **diffusion.** This is what happens, for instance, when a lump of sugar dissolves in a cup of coffee. The warmer the liquid, the more collisions, hence the faster the diffusion. The rate of diffusion also depends upon the size of the molecules. Larger molecules diffuse more slowly than smaller ones, because it takes more of a shove to get them moving.

The process of diffusion is exceedingly important to life. The high concentration of organic molecules within cells is very unlikely. Without the cell membrane these organic molecules would diffuse outward and become rapidly unavailable for the cell's activities. Thus the membrane holds them in. Yet the membrane must not act a barrier to all molecules, because important new molecules (e.g., food) must enter the cell and a variety of cell products, including waste molecules, must be able to exit. Cell membranes are said to be **selectively permeable,** as they, in fact, allow some molecules to pass through while others cannot.

Osmosis

Suppose a cell that contains water molecules plus a number of large organic molecules is placed in a solution of pure (i.e., distilled) water. Suppose further that the cell has a selectively permeable membrane that allows the smaller water molecules to pass through but not the larger organic molecules. What will happen to the cell? Because the concentration of water molecules on the inside of the cell is lower than on the outside (see Figure 10.5), a greater proportion of water molecules will be passing in than out. This is because there are fewer water molecules on the inside to pass out than on the outside to pass in. Also, those on the inside have a greater chance of hitting a large organic molecule and bouncing back toward the inside than those on the outside. Thus, water will tend to diffuse into the cell. (A net movement of a solvent, usually water, through a selectively permeable membrane is called **osmosis.**) Because the larger molecules cannot diffuse out, the net concentration of molecules inside the cell will increase, and the cell will swell and may burst. When a cell is placed in a solution in which the water concentration is greater inside the cell, more water will move out than in, and the cell will shrink. Most cell membranes are relatively permeable to water but are relatively impermeable to larger molecules such as starches and proteins.

Figure 10.6 shows a generally accepted "fluid mosaic" model of a cell membrane that can account for these observations. In the model membrane, a double layer of lipids form the main part of the mem-

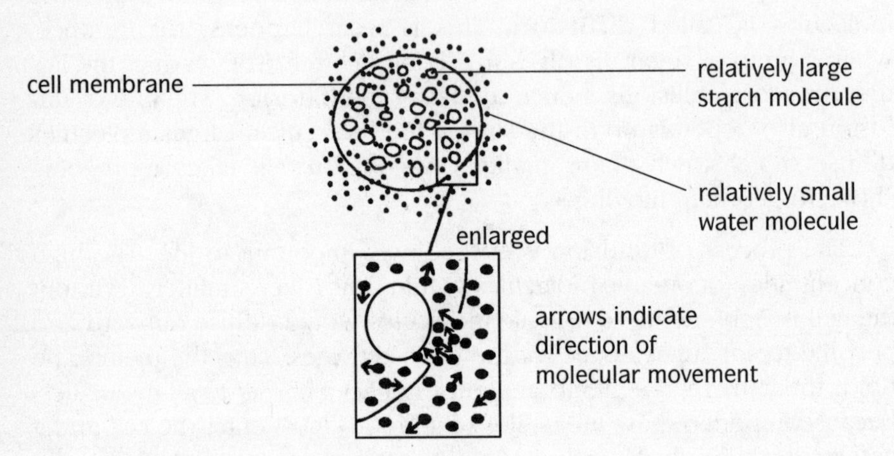

Figure 10.5 Representation of a cell membrane showing the net movement of water molecules into the cell.

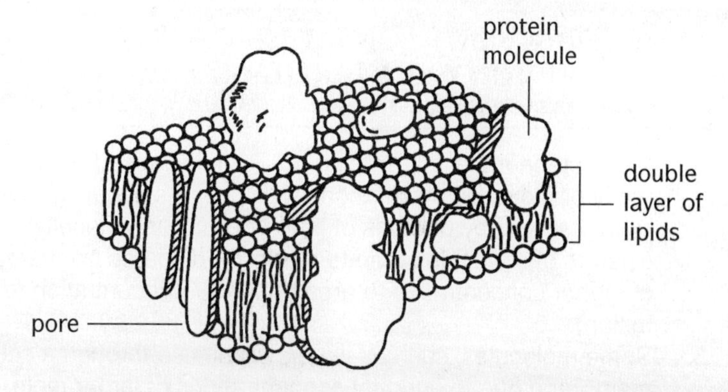

Figure 10.6 The fluid-mosaic model of a cell membrane.

brane. Proteins are embedded in the lipid layers. Membrane pores occur as channels through one or a group of protein molecules. The pores are too small to allow passage of large molecules.

Active Transport

Not all molecules pass through the cell membrane due to the random and passive forces of diffusion and osmosis. The cell membrane itself is capable of actively causing some molecules to pass through that are apparently too large to fit through the tiny holes in the membrane. The membrane also is able to force some molecules through from an area of lower to higher concentrations. In each case, however, energy must be expended to do so. This form of passage through the membrane is called **active transport.** The specific way(s) the membranes accomplishes active transport are not yet fully understood.

POSTULATES OF THE THEORY OF MOLECULAR MOVEMENT

1. Much of the molecular motion on Earth is caused either directly or indirectly by collisions with photons of sunlight.
2. When "struck" by photons of light or by other molecules, clusters of molecules tend to randomly disperse from areas of higher concentration to areas of lower concentration (diffusion).
3. Some molecules, such as water, will diffuse through a cell membrane due to unequal concentrations of larger molecules that are themselves unable to diffuse through the membrane (osmosis).
4. The cell membrane is capable of expending energy to force some relatively large molecules through and to force some molecules to move from areas of lower to higher concentration (active transport).

QUESTIONS

1. Do you know of other reasoning and evidence that supports the atomic hypothesis? If so what is it?

____

____

____

____

2. Is the hypothesis that atoms are indivisible still accepted today? If not, what evidence exists that does not support this hypothesis?

____

____

____

____

3. The fact that your body stays at about 99°F, even on cold days, suggests that most of the chemical reactions in your cells are exothermic. Where do you get the large molecules involved in these reactions? What does your body ultimately do with the smaller molecules that result from the breakdown process?

4. Postulate 5 of the metabolism theory states that each chemical reaction is governed by an enzyme specific to that reaction. Suppose you know that a series of reactions convert chemical A to B, B to C, and C to D. You also know that this series of reactions requires three enzymes numbered 1, 2, and 3. How could you research to discover which of the reactions is governed by each of the enzymes?

5. What would happen to the size of red blood cells placed in distilled water? In water that has a concentration of other types of molecules much higher than that found in the red blood cells? In water that has a concentration of other types of molecules equal to that found in the red blood cells? Explain.

6. Given a normal cell sitting in distilled water, the distilled water will move into the cell. This movement is called osmosis. The standard explanation for osmosis is that the water molecules

move from an area of high concentration of water molecules outside the cell to an area of low concentration of water molecules inside the cell. What forces are presumably responsible for this osmotic movement?

7. If it is true that water molecules move into the cell mentioned above due to collisions with other molecules, why don't the high concentrations of other types of molecules already inside the cell prevent the water molecules from entering? Most likely you do not already "know" the answer to this question. Nevertheless, you might be able to suggest some possible answers.

TERMS TO KNOW

activation energy
active transport
amino acid
carbohydrate
catalyst
cell membrane
chemical reaction
diffusion
endothermic
energy
enzyme
exothermic
isotope
kinetic energy
lipid
metabolic rate
metabolism
nucleic acid
nucleotide
osmosis
pH (scale) (potential of hydrogen)
potential energy
protein
selectively permeable
temperature

SAMPLE EXAM

Circle or enter the letter of the best answer (e.g., a, b, c, d). In Table 10.3, a plus sign (+) indicates a positive reaction, zero (0) indicates no reaction, and a blank indicates that no test was made.

Use the data in the table and the following key to answer questions 1–3.

KEY: a. Statement is probably true.

b. Statement is probably false.

c. Insufficient data.

TABLE 10.3 REACTIONS OF UNKNOWN SUBSTANCES WITH SEVERAL INDICATORS

	INDICATOR					
UNKNOWN SUBSTANCE	IODINE (a)	BENEDICT'S (b)	PAPER (c)	BIURET (d)	BTB (e)	PHENOL-PHTHALEIN (f)
1	+	0	0	0		
2	+	+		0		
3	+					red
4	+		+		yellow	
5	+				yellow	
6		0	+	0	blue	red
7	0	0	+	+		

(a) tests for starch; (b) tests for sugars; (c) tests for fats; (d) tests for proteins; (e) blue → green → yellow indicates CO_2; (f) colorless → red indicates base.

1. Unknown 2 contains starch molecules, some of which have been broken apart into smaller molecules. ____________

2. Unknown 6 has a pH greater than 7 and contains starch and fat molecules. ____________

3. Unknown 7 is most likely a potato. ____________

The next four questions (4–7) refer to the following experimental results. Four experiments were conducted in four separate test tubes. Bubbles were seen in some of the test tubes, as indicated in Table 10.4.

TABLE 10.4 EXPERIMENTAL RESULTS

TEST TUBE NO.	CONTENTS OF TEST TUBES	RESULTS
1	new liver + new hydrogen peroxide	bubbles
2	old liver + new hydrogen peroxide	bubbles
3	old liver + old hydrogen peroxide	no bubbles
4	new liver + old hydrogen peroxide	no bubbles

4. Suppose the hypothesis is advanced that molecules in the liver change but those in hydrogen peroxide do not change when the two substances are mixed. What predicted result can be derived from this hypothesis and experiment? The mixture in

   a. test tube 3 should produce bubbles.

   b. test tube 1 should not produce bubbles.

   c. test tube 4 should produce bubbles.

   d. test tube 4 should not produce bubbles.

   e. test tube 2 should produce bubbles.

5. Since bubbles were produced only in test tubes 1 and 2, the hypothesis in number 4 has

   a. been supported.

   b. been contradicted.

   c. not been tested.

6. Suppose the alternative hypothesis is advanced that molecules in the liver do not change but that those in hydrogen peroxide do

change when the two substances are mixed. What predicted result can be derived from this hypothesis and experiment? The mixture in

a. test tube 1 should not produce bubbles.

b. test tube 2 should produce bubbles.

c. test tube 2 should not produce bubbles.

d. test tube 3 should produce bubbles.

e. test tube 4 should produce bubbles.

7. A biologist dilutes blood cells with water on a glass slide and observes them through a microscope. The cells seem to burst. This is probably because he

a. used distilled water.

b. used very salty water.

c. added the water too rapidly.

d. used dead cells.

For the next two items (8 and 9), use the diagram shown in Figure 10.7.

8. The cell in Figure 10.7 would probably

a. expand and burst open.

b. shrivel up.

c. retain its normal shape.

d. lose its salt content.

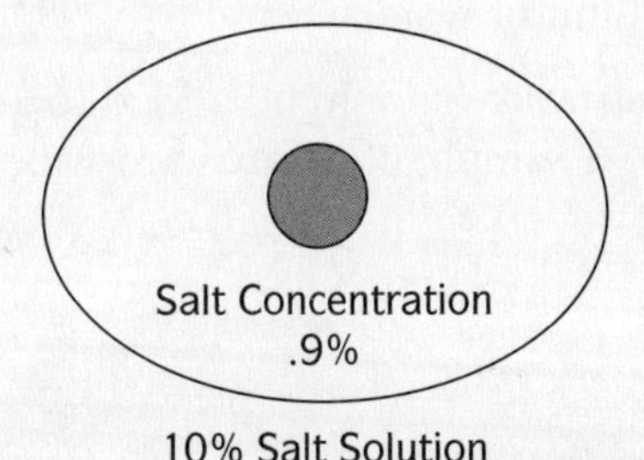

Figure 10.7 The salt concentration inside and outside a cell.

9. The concentration of water is
   a. greater inside the cell than outside.
   b. greater in the immediate environment than in the cell.
   c. equal inside and outside the cell.
   d. not important to the size of the cell.
10. Assuming that diffusion of molecules takes place through cell membranes, which of the following cells is more efficient at allowing molecules in and out of cell?
    a. 1 cm on a side
    b. 2 cms on a side
    c. 3 cms on a side
    d. 4 cms on a side
    e. 5 cms on a side
11. Which of the following factors is the most important in limiting the size of cells?
    a. density
    b. surface area
    c. volume
    d. weight
    e. surface area/volume

Use the following information to answer questions 12 and 13.

Suppose you are given a sample of salt water with five times the salt found in normal sea water. Normal sea water contains about 35 grams of salt per 1000 ml of water.

12. How much fresh water should be added to 10 ml of the 5X solution to produce a sample of normal sea water?
    a. 10 ml
    b. 20 ml
    c. 30 ml
    d. 40 ml
    e. 50 ml

13. How much fresh water should be added to 10 ml of the 5X solution to produce a sample of water that has half the salt concentration of normal sea water?

    a. 10 ml

    b. 20 ml

    c. 50 ml

    d. 90 ml

    e. 100 ml

CHAPTER 11

SPECIFIC MOLECULAR LEVEL THEORIES

GETTING FOCUSED

- *Identify a theory of cellular fermentation and respiration.*
- *Learn about a theory of photosynthesis.*
- *Learn about Watson and Crick's theory of DNA structure and replication.*
- *Evaluate a theory of gene function.*

A THEORY OF CELLULAR FERMENTATION AND RESPIRATION

Just how molecules interact with one another inside cells is a topic of considerable research and continued mystery. In general, the interactions involve either breaking molecules apart, rearranging the atoms in molecules, constructing larger molecules from smaller ones, moving molecules from one place to another, or transferring electrons from one molecule to another. The problem is to figure out how these interactions, taken together, accomplish the cellular activities of movement, growth, replication, maintenance, and repair.

A number of theories have been generated that attempt to describe and explain these interactions. One such theory attempts to explain how relatively large, slow-moving food molecules such as carbohydrates, proteins, and fats are broken into smaller, fast-moving molecules to provide motion (energy) for the cell's activities.

Interestingly, it has been discovered that cells are able to break relatively large, slow-moving food molecules apart to form small, fast-moving molecules of carbon dioxide and water only when a sufficient number of oxygen molecules are present. This process is referred to as **cellular respiration.** When oxygen molecules are not present, the food molecules are only partially broken apart. This partial breakdown is called **fermentation.** Fermentation provides a lot less usable energy for the cell than does respiration.

Providing Energy

Cellular respiration and fermentation are processes in which relatively large food molecules are broken into smaller molecules to provide energy for the cell's activities. But just what does the phrase "provide energy" mean? This is not an easy question to answer. Perhaps the use of an analogy will help.

Have you ever dropped a glass bowl on a hard floor? When the bowl hits the floor, it shatters and tiny pieces of glass fly off in all directions, sometimes traveling several yards before hitting something and stopping. A piece of flying glass might even cut you. In a somewhat similar way, cells are able to take in relatively large food molecules (the glass bowl) and break them apart (the bowl strikes the floor) to release the motion/energy bound up in their atomic bonds. When the bonds are broken, the now rapidly moving atoms (flying pieces of glass) can strike other atoms and/or molecules and make them move.

The motion is transferred to the object that gets hit, so that the once rapidly moving atoms slow down or stop moving. This transfer of motion/energy to the other atoms and/or molecules causes them to move, and possibly stick together with other atoms and/or molecules, or perhaps break apart (flying glass cutting your leg). In other words, when the phrase "providing energy" is used, it means that the motion stored up in the bonds of molecules (chemical energy) is released in the form of rapidly moving electrons, atoms, and/or molecules (kinetic energy) that can now transfer their motion/energy to other atoms and/or molecules by striking them.

A major difference between breaking a glass bowl by dropping it on the floor and breaking a food molecule is, of course, one of size. The bowl is a lot bigger than a food molecule. Another important difference is the rate of breakdown. The bowl shatters all at once, and the motion of its flying pieces does nothing useful. It can even do damage. In the cell, however, the food molecules are broken apart piece by piece so that the faster-moving pieces that move off can be directed to accomplish useful work. For example, one very useful job is for the moving pieces to strike nearby atoms to get them moving so that they strike and stick onto (bond with) other molecules to form still larger molecules. The new, larger molecules can be transported about and can in turn be broken apart to provide energy/motion for still other jobs. This is, in fact, the central job of cellular fermentation and respiration.

Cellular Fermentation

As mentioned, the process of fermentation takes place without oxygen being present. When the food molecule is glucose (a type of sugar), the first step in fermentation involves breaking glucose apart to form two molecules called pyruvic acid ($C_3H_4O_3$). This is also the first step in cellular respiration and is called *glycolysis*. In the process two molecules called **adenosine diphosphate** interact and bond with two molecular fragments called phosphates to create two larger molecules called **adenosine triphosphate** (ATP for short). The general equation for the fermentation of glucose looks like this:

$$1\ C_6H_{12}O_6 + 2\ \text{ADP's} + 2\ \text{phosphate fragments} \rightarrow 2\ C_3H_4O_3 + 2\ \text{ATP's}$$

(glucose)

The choice of a glucose molecule as an example is intentional. Many large food molecules are first broken down to smaller glucose molecules before they are further broken down.

Cellular Respiration

When an excess of oxygen (O_2) is present, the cell is able to break apart a molecule, such as glucose ($C_6H_{12}O_6$), to yield six smaller molecules of carbon dioxide (CO_2) and six water molecules (H_2O). In the process, 38 molecules of adenosine diphosphate interact and bond with 38 phosphate molecular fragments to create 38 larger molecules of adenosine triphosphate.

The general equation for cellular respiration of a molecule of glucose looks like this:

$$1\ C_6H_{12}O_6 + 6\ O_2 + 38\ \text{ADP's} + 38\ \text{phosphate fragments} \rightarrow 6\ CO_2 + 6\ H_2O + 38\ \text{ATP's}$$

The equations for cellular fermentation and respiration show only the molecules present at the start and then at the end of the breakdown process. Biochemists have conducted considerable research to try to determine precisely what intermediate molecular interactions take place during the breakdown process. A great number of these intermediate interactions are now known. Although we need not concern ourselves with the details of these intermediate interactions, the basic sequence of interactions is shown in Figure 11.1. To keep matters relatively simple, only the carbon atoms are shown, but keep in mind that the intermediate molecules also contain oxygen and hydrogen atoms (amino acids also contain nitrogen atoms).

The process begins in the digestive tract, where relatively large food molecules such as carbohydrates and proteins are first broken into smaller glucose and amino acid molecules. Next, the glucose and amino acid molecules are broken apart to produce still smaller molecules with only three atoms of carbon (pyruvic acid). As shown in Figure 11.1, these three-carbon pyruvic acid molecules then enter the mitochondria, where they are broken apart to produce a two-carbon molecule (called acetic acid) and a single carbon atom. Note that fat molecules are broken down to produce a two-carbon molecule as well.

Next, the two-carbon molecule combines with a four-carbon molecule to produce a six-carbon molecule (called citric acid). The six-carbon molecule is then broken apart to form a five-carbon molecule and a single carbon atom. Two hydrogen atoms also are broken off the molecule. Next, one more carbon atom and two more hydrogen atoms are broken off to leave a four-carbon molecule. Four more hydrogen atoms then break away from the four-carbon molecule, and it is ready to join another two-carbon molecule to start the cycle all over again.

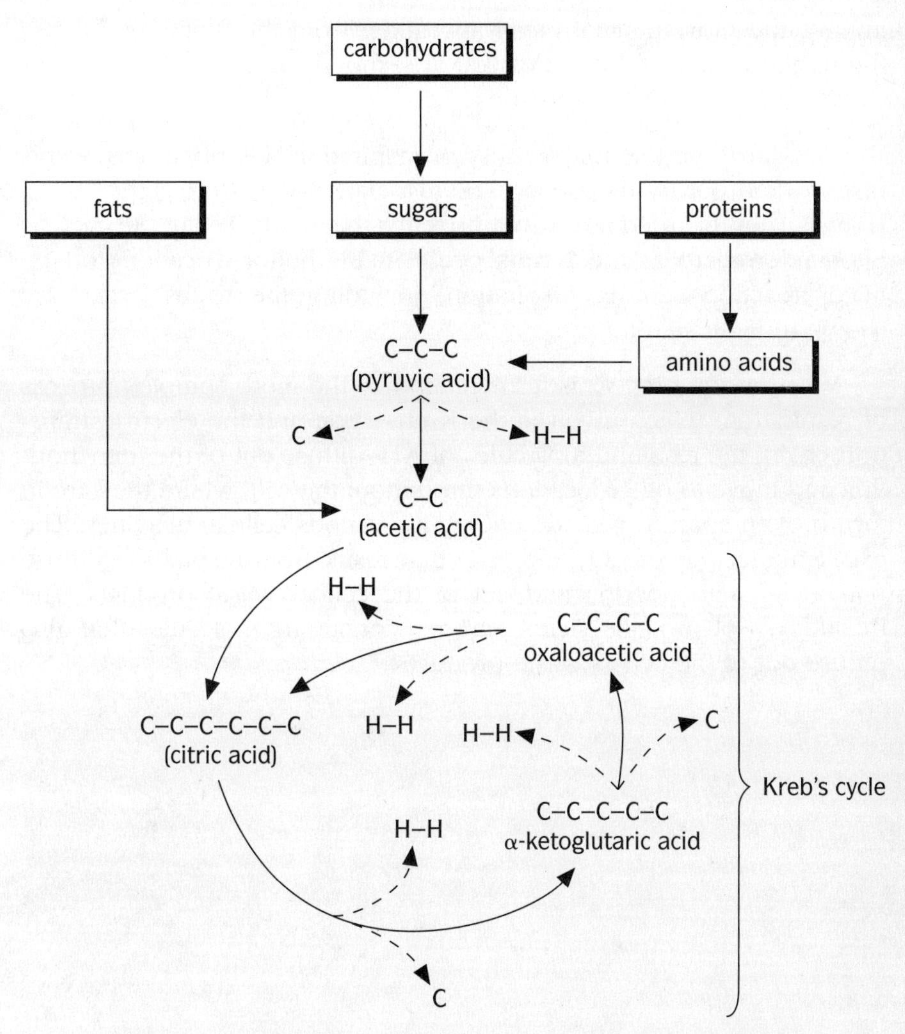

Figure 11.1 The pattern of cellular fermentation and respiration.

The single carbon atoms that break away from the chains of carbon atoms eventually combine with O_2 molecules to form carbon dioxide molecules. The hydrogen atoms that break away also join O_2 molecules. This, of course, forms H_2O molecules (water). The motion of the hydrogen atoms indirectly provides the force used to cause ADP molecules and the phosphate fragments to strike one another and combine to produce ATP molecules. The details of this indirect and complex set of interactions have been partially worked out. The interactions are collectively referred to as the **electron transport chain** because they involve the movement of electrons (taken from the hydrogen atoms) through a "chain" of molecules that are embedded next to one another in the mitochondrion membrane.

The cyclic part of the process of respiration described above was first worked out by the British scientist Hans Krebs during the 1930s. He was later awarded the Nobel prize for this work. Today the cycle is often referred to as the **Krebs cycle** in his honor. Locations of the major reactions during respiration, including the Krebs cycle, are shown in Figure 11.2.

Whether the process is fermentation or the more complex process of respiration, which involves the Krebs cycle and the electron transport chain, the resulting molecules of ATP diffuse out of the mitochondria and move to other locations throughout the cell, where they are in turn broken apart to provide energy for various cellular activities. The excess molecules of CO_2 and H_2O that result from respiration diffuse out of the mitochondria and out of the cell as waste products. The breakdown of proteins yields nitrogen-containing molecules that also diffuse out of the cell as waste products.

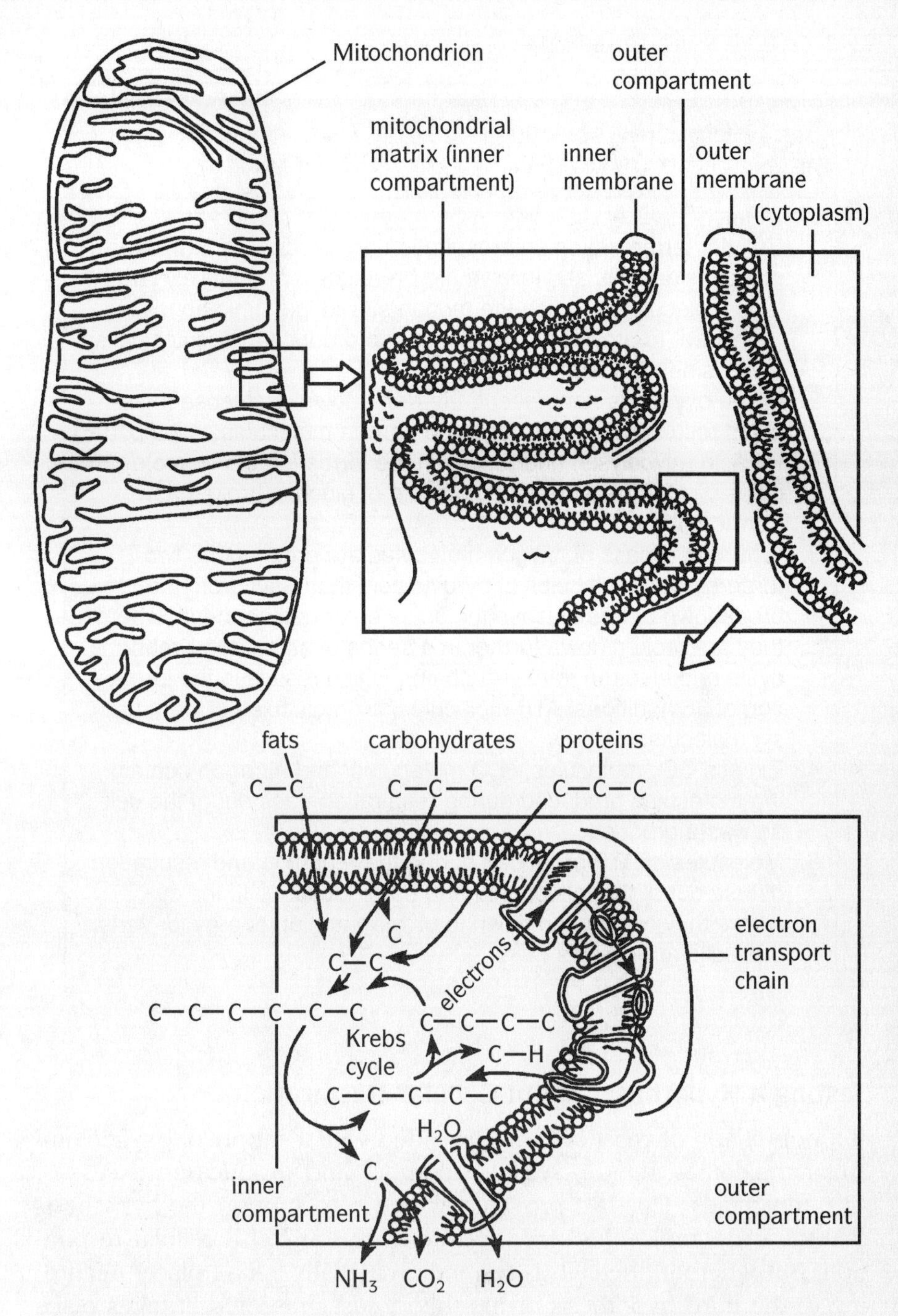

Figure 11.2 Progressively more detailed views of a mitochondrion showing its major structures and the hypothesized locations of key events during cellular respiration.

POSTULATES OF THE THEORY OF CELLULAR FERMENTATION AND RESPIRATION

1. Cellular fermentation and respiration are processes that take place inside cells and involve the breakdown of relatively large food molecules in which the motion stored in atomic bonds is "released" to provide energy/motion to produce molecules called adenosine triphosphate (ATP).
2. When oxygen is not present, the breakdown of glucose molecules results in two smaller three-carbon molecules called pyruvic acid (glycolysis) and results in the formation of two molecules of ATP for every one molecule of glucose broken down (fermentation).
3. When an excess of oxygen molecules are present, the two three-carbon molecules of pyruvic acid that result from the breakdown of one glucose molecule enter mitochondria, where they are broken down further in a series of steps (the Krebs cycle and electron transport chain), which results in the formation of 36 additional ATP molecules, six molecules of CO_2, and six molecules of H_2O.
4. Excess CO_2 molecules, H_2O molecules, and nitrogen-containing molecules produced during respiration pass out of the cell as waste products.
5. Molecules of ATP produced during fermentation and respiration diffuse out of the mitochondria to other locations in the cell, where they are broken down to provide motion/energy for various cellular activities.

Testing a Hypothesis About Cellular Respiration

If a population of yeast cells are provided with a supply of oxygen and sugar molecules, they increase in number and produce an excess of CO_2 molecules. This fact provides evidence to support the hypothesis that the sugar molecules serve as an energy source for cellular respiration. But where in the cells is respiration actually taking place? According to the third hypothesis of the theory just presented, it takes place inside the mitochondria. How can this hypothesis be tested?

One way involves separating the mitochondria from the other cellular organelles. This can be accomplished by putting a batch of cells (liver cells, kidney cells, leaf cells) in a blender and blending them to break up their cell membranes and/or cell walls. The cell contents can then be poured into a test tube and spun rapidly in a machine called a centrifuge. Because the cell organelles have different densities, the centrifugal force created by spinning the test tube around and around will cause the organelles to move to different layers in the test tube. The most dense organelles will settle to the bottom of the tube, while the least dense organelles will stay near the top (see Figure 11.3). Now the layers of organelles can be poured into separate test tubes and the hypothesis that respiration takes place inside mitochondria can be tested as follows:

Hypothesis: *If*...respiration takes place inside mitochondria...

and...mitochondria are suspended in a solution that contains glucose and oxygen...

Prediction: *then*...carbon dioxide should be produced.

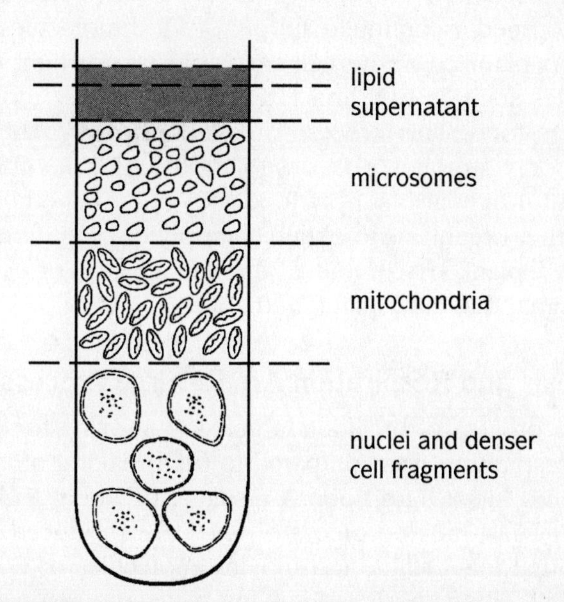

Figure 11.3 Test tube showing the layers of organelles after being spun in a centrifuge.

On the other hand,

Hypothesis: *If*...respiration takes place in some other organelles, such as the cell nucleus...

and...nuclei are suspended in a solution that contains glucose and oxygen...

then...carbon dioxide should be produced.

When such experiments are performed, the results show that carbon dioxide is produced only in the test tubes containing mitochondria. Therefore, the hypothesis that respiration takes place inside mitochondria has been supported, and the hypothesis that it takes place in other types of organelles has not been supported.

A THEORY OF PHOTOSYNTHESIS

All living cells use relatively large molecules to provide the energy needed to sustain life. In general, the energy is provided when the large molecules are broken apart during the complex processes of fermentation and respiration. Some cells, such as animal cells, need a continued supply of food molecules to stay alive. When in the dark, plant cells also need a continued supply of food molecules. But when light shines on plant cells, the cells are able to use light energy and small molecules such as CO_2 and H_2O to make larger molecules that can be used as food. This process is called **photosynthesis.** Photosynthesis not only supplies plants with food molecules for their own respiration, but it may also supply food molecules for other organisms as well. For other organisms to obtain a supply of food molecules, they need only eat a plant, eat an animal that ate a plant, or eat an animal that ate an animal that ate a plant, and so on.

Photosynthesis and Respiration Are Reverse Processes

Recall that the process of breaking large, slow-moving food molecules apart during respiration was compared to the breaking of a glass bowl when it falls and hits a hard floor. A result of breaking a glass bowl is that lots of little fast-moving pieces of glass are produced. A result of respiration is that lots of little fast-moving molecules of CO_2 and H_2O are produced. Photosynthesis is, in many ways, the reverse of respiration. A main task of photosynthesis is to take little molecules of CO_2 and get them to combine with one another to create larger molecules.

Just as trying to sweep up all the broken pieces of glass and sticking them together requires energy, so does collecting the CO_2 molecules and sticking them together. Whereas you supply the energy/motion to sweep up the broken glass, light supplies the energy/motion to drive photosynthesis.

What Is the Nature of Light?

Light is a very strange thing. To date, no one has generated a completely satisfactory idea of just what light is. In some ways, light appears to be like streams of tiny particles called **photons** that travel through space in straight lines. In other ways, light appears to consist of waves with distinct wave lengths, but waves of what?

Let us consider in more detail the notion that light consists of particles. The discussion will reveal some of the complexities and problems with our current ideas of just what light is. Suppose you place a tiny fan with blades painted black on one side and silver on the other in a jar partially filled with air. When a light shines on the black side of the blades, the blades will begin to rotate towards the silver sides. This observation can be interpreted by imagining that light consists of tiny particles that actually strike the blades much like tiny bullets that hit and turn the blades. However, if light consists of tiny particles, then they would have to go somewhere after they strike the blades. Where? By analogy with the bullet idea, you might guess that they would glance off the blades and continue on their way in some new direction with a little less speed. This idea fails on two counts. First, when the speed of light in air is measured, it always turns out to be the same, no matter how many objects it glances off of. Second, there is no evidence that light even glances off black objects such as the fan blades. Instead, light appears to be absorbed by black objects.

Therefore, it may be better to interpret the results of the fan experiment as an indication that the particles get absorbed by the blades, that they somehow get stuck in the blades. If this is so, then one would expect that the blades would get heavier after light shines on them. But they do not! So what happened to the particles?

Before we attempt to answer this question, let us consider another experiment. When light shines for a few minutes on a test tube of chlorophyll molecules suspended in water and then is turned off, the chlorophyll molecules glow. This observation suggests that the chlorophyll molecules, like the black fan blades, absorb some of the photons when the light is on and then emit them when the light is off. This idea, coupled with the observation that objects that absorb photons do not gain weight, suggests that (1) either photons are weightless particles (a difficult idea to swallow), (2) they cease to exist when they are absorbed only to be recreated when they are emitted (an equally difficult idea to swallow), or (3) both statements one and two are correct. Indeed, this third possibility is the way many scientists have come to think about photons that, although entirely correct mathematically, leaves a lot to be desired conceptually.

Thus, the answer to the question What happened to the particles after they hit the fan blades? would go something like this: The weightless photons "interact" (in some unknown way) with the electrons of the atoms in the blades to make them move more rapidly. (That is, the energy/motion of the photons is transferred to the electrons, somewhat like the way a football picks up energy/motion when it is kicked by a foot, but in this case the electrons are getting hit by a disappearing, weightless foot.) Hence, the black side of the blades heats up. (Remember that temperature is a measure of the amount of motion in the atoms/molecules of a substance.) Some of this increased motion in the black side of the blades is then transferred to nearby air molecules, which bounce away from the black surfaces, creating a force that propels the blades in the opposite direction. The fan turns.

In summary, an answer to the question What is the nature of light? is as follows: Light behaves at times as though it consists of waves of some unknown something, and at other times as though it consists of streams of tiny, weightless, disappearing, and reappearing particles that are nevertheless able to transfer energy/motion to objects by somehow imparting energy/motion to the electrons of those objects.

How Does Photosynthesis Take Place?

The Light-Dependent Reactions Plant cells contain organelles called **chloroplasts.** Inside the chloroplasts are tiny disc-like structures called **thylakoids** that are stacked on top of one another like pancakes. Individual thylakoids consist of a double-layered membrane that surrounds a water-filled cavity. Large molecules of various kinds are embedded next to one another in the thylakoid membranes. Some of these are chlorophyll molecules (see Figure 11.4).

When light (of a specific wavelength) shines on the chlorophyll molecules, some of the photons interact with electrons in the chlorophyll

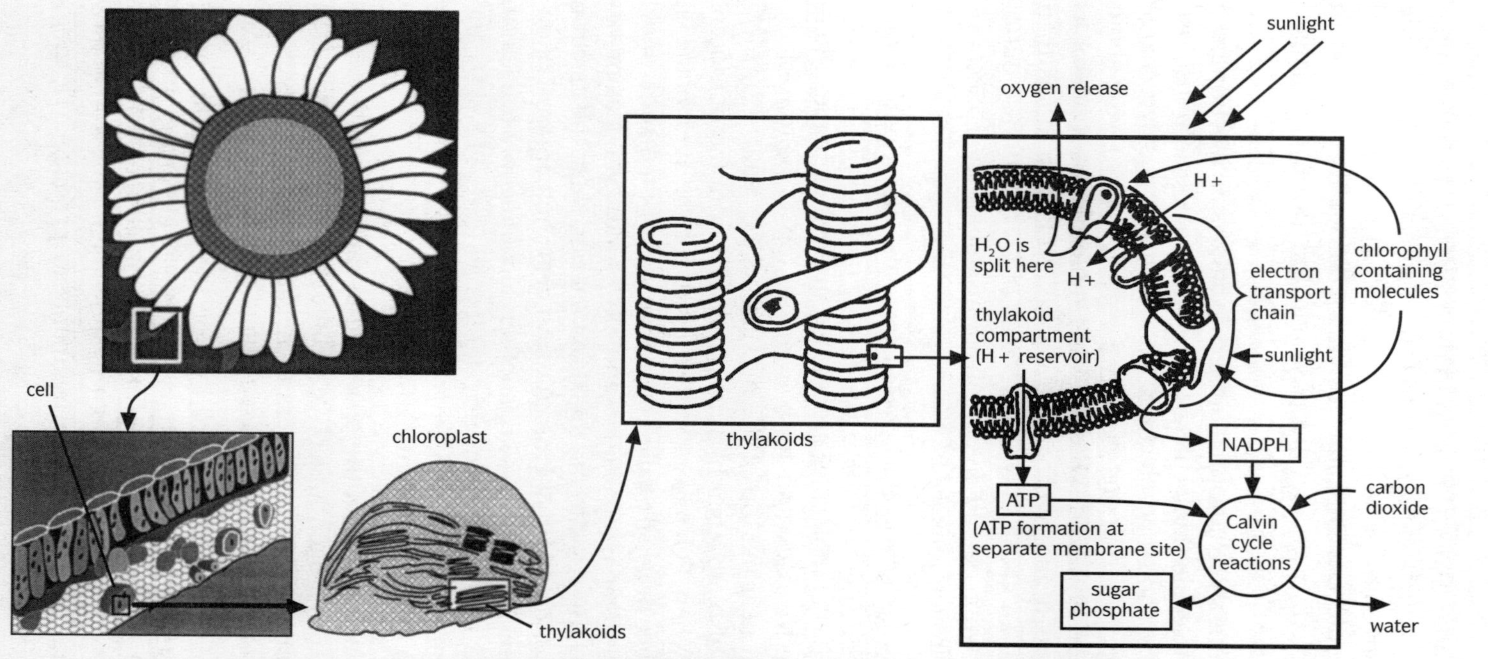

Figure 11.4 A section through a leaf of a sunflower plant showing progressively more detailed views of the cells of the leaf. The more detailed view is of a single chloroplast within a single cell. Note the stacks of disk-like structures called *thylakoids.* A stack of thylakoids is called a *granum.* Stacks of thylakoids are shown in the center. The enlarged diagram at the right shows a section of one thylakoid membrane. The light-dependent reactions of photosynthesis presumably take place on the thylakoid membranes as shown. The light-independent reactions presumably take place outside the thylakoids, but still inside the chloroplasts.

molecules to impart additional motion to the electrons. Due to its negative charge, this newly "energized" electron is attracted to a molecule with a slight positive charge that is located next to it in the thylakoid membrane (molecule Q). This attractive force causes the electron to leave the chlorophyll molecule and pass into this positively charged molecule.

In the meantime, the chlorophyll molecule attracts an electron from another nearby molecule (molecule Z) to replace the electron just lost. Molecule Z is now short one electron. This creates a positive charge that attracts an electron from one of the water molecules moving about on the inside of the thylakoid. When a water molecule loses one of its electrons, the oxygen and hydrogen atoms split. Oxygen atoms then combine with one another to form molecules of O_2 that diffuse out of the thylakoid, leaving behind hydrogen atoms minus their electrons (i.e., protons, hydrogen ions, H^+).

The "energized" electron that passed from the chlorophyll molecule to molecule Q quickly passes to another molecule in the membrane (PQ). Lots of PQ molecules are located in the membrane, where they randomly move back and forth from the top to bottom of the membrane. When near the top of the membrane, PQ attracts the electron from Q and also attracts a hydrogen ion (H^+) from the outside of the membrane. When PQ moves to the inside of the membrane, it moves near another molecule called cyt f. Because of the nature of cyt f, the energized electron is attracted to and moves into cyt f, leaving behind the H^+, which passes into the thylakoid interior. Next, the electron passes into another chlorophyll molecule and gets another boost of energy/motion when it interacts with another incoming photon. From there it is passed to still other nearby molecules (i.e., FES, FD, and FAD), until it ultimately joins with a hydrogen ion (H^+) and a molecule called NADP to produce a molecule of NADPH.

The major consequence of the process thus far is that an excess of positively charged hydrogen ions (H^+) accumulate on the inside of the thylakoid membrane, while an excess of negatively charged hydroxyl ions (OH^-) and NADPH molecules accumulate on the outside.

The excess of negatively charged ions on the outside of the membrane sets up a strong force that attracts the positively charged hydrogen ions through a permeable group of protein molecules (CR_1) to the outside. In a way not yet understood, this movement causes ADP molecules to combine with phosphate fragments to produce ATP molecules that accumulate on the outside. Because all of the reactions just described require light, they are collectively referred to as the **light-dependent reactions** of photosynthesis.

The Light-Independent Reactions The light-dependent reactions of photosynthesis result in the accumulation of two types of molecules (ATP and NADPH) on the outside of the thylakoid discs. These molecules are now used to "drive" the next series of reactions, which result in the synthesis of large molecules (such as sugars and amino acids) from small molecules (such as CO_2). This synthesis takes place outside the thylakoids but still inside the chloroplasts in a series of cyclic reactions referred to as the **Calvin cycle.** The Calvin cycle is named after the American scientist, Melvin Calvin, who discovered its steps in a series of experiments conducted during the 1940s. Because the reactions of the Calvin cycle do not require light, they are collectively referred to as the **light-independent reactions** of photosynthesis.

The Calvin cycle begins when a carbon dioxide molecule combines with a five-carbon molecule to form an unstable six-carbon molecule that quickly breaks apart to form two three-carbon molecules. These three-carbon molecules then combine with H^+ ions and phosphate fragments to form stable three-carbon molecules. Many of these stable three-carbon molecules accumulate in this way. Most of them combine with two-carbon molecules to produce five-carbon molecules that are then available to join another CO_2 molecule to start the cycle again. The remainder combine with one another to produce the sugars that are considered to be the primary products of photosynthesis, and later starches, amino acids, lipids, and nucleotides.

Some of the complex molecules synthesized by the light-independent reactions move out of the chloroplasts to be used at once in plant respiration or for other necessary activities. Excess molecules are stored to be used later by the plant or are used by other organisms that eat plants.

POSTULATES OF PHOTOSYNTHESIS THEORY

1. Photosynthesis consists of two sets of reactions—one set requires light, the light-dependent reactions; the other set does not require light, the light-independent reactions.
2. The light-dependent reactions occur in the thylakoids of the chloroplasts when photons interact with electrons of chlorophyll molecules and cause them to move more rapidly.
3. The light-dependent reactions break apart H_2O molecules to produce O_2 molecules and H^+ ions, and in the process convert ADP to ATP and NADP to NADPH.
4. The light-independent reactions occur in the chloroplasts outside the thylakoids.
5. The light-independent reactions (including the Calvin cycle) combine several CO_2 molecules and several H^+ ions to produce large molecules, such as glucose ($C_6H_{12}O_6$), and in the process reconverts ATP to ADP and NADPH to NADP.
6. Most of the CO_2 used in the light-independent reactions comes from the air, but some comes from cellular respiration inside the plant cell.
7. Most of the O_2 produced in the light-dependent reactions goes into the air, but some is used inside the plant cell in cellular respiration.

Testing Postulates of the Theory

Where do the molecules of oxygen that are produced during photosynthesis come from? A simplified equation for the overall process of photosynthesis looks like this:

$$6\ CO_2 + 6\ H_2O + \text{light} \rightarrow 1\ C_6H_{12}O_6 + 6\ O_2$$

The equation tells us that one molecule of the sugar glucose (1 $C_6H_{12}O_6$) and six molecules of O_2 (6 O_2) are produced for every six molecules of CO_2 (6 CO_2) and six molecules of H_2O (6 H_2O) used. Thus, according to the equation, the O_2 could come from one of two types of molecules. It could come from the CO_2 molecules or it could come from the H_2O molecules. Which is it?

The third and the seventh postulates of the theory just presented claim that the O_2 comes from the splitting of water molecules. This is consistent with the claim in postulate 5 that the CO_2 molecules are used to produce glucose. What evidence exists to support these claims?

In 1941 a group of American scientists performed two experiments to determine the source of the O_2. The experiments involved green algae living in water. The water had been labeled with an isotope of oxygen called O^{18}, which can be detected with a machine called a mass spectrophotometer. The reasoning that guided their first experiment follows:

Hypothesis: *If*...the oxygen produced during photosynthesis comes from water molecules...

and...the water molecules contain the O^{18} isotope...

Prediction: *then*...the oxygen produced by the green algae should contain the O^{18} isotope.

On the other hand,

Hypothesis: *If*...the oxygen produced during photosynthesis comes from carbon dioxide molecules...

and...the water molecules contain the O^{18} isotope...

Prediction: *then*...the oxygen produced by the green algae should not contain the O^{18} isotope.

Results of the experiment were that all the oxygen produced contained the O^{18} isotope, just as predicted by the first hypothesis. Thus, the result supports the hypothesis that the oxygen comes from splitting water molecules, and not from splitting CO_2 molecules. Just to be certain of this conclusion, a second experiment was performed. In the second experiment, the CO_2 molecules were labeled with the O^{18} isotope instead of the H_2O molecules. In this case, the predictions were that the O^{18} would show up in the glucose molecules and not in the oxygen molecules. The results turned out precisely as predicted. Thus, we can be fairly certain that the O_2 comes from the H_2O molecules and not from the CO_2 molecules.

WATSON AND CRICK'S THEORY OF DNA STRUCTURE AND REPLICATION

In 1962 James Watson, Francis Crick, and Maurice Wilkens were awarded the Nobel prize for their research on determination of the molecular structure of DNA, the molecule contained in chromosomes and responsible for transmitting genetic information from parent to offspring. The paper that detailed the theory was entitled "Molecular Structure of Nucleic Acids: A Structure for Deoxyribose Nucleic Acid." It was written by Watson and Crick and had appeared nearly ten years earlier, in 1953, in the British journal *Nature*.

POSTULATES OF WATSON AND CRICK'S THEORY OF DNA STRUCTURE AND REPLICATION

1. DNA consists of a two-stranded spiral helix with the two strands of the spiral helix made up of alternate molecules of deoxyribose and phosphoric acid.
2. Pairs of base molecules form links between the two opposite deoxyribose molecules in the two strands. The base pairs consist of adenine-thymine combinations (-A-T-) and cytosine-guanine (-C-G-) combinations.
3. The base pairs may be in any sequence along the two-stranded spiral helix and may be positioned -A-T-, -T-A-, -C-G-, or -G-C-.
4. When DNA replicates, the two bases in each pair separate, permitting the two strands to separate.
5. As the strands separate, complementary nucleotides pair sequentially with the bases in each strand.
6. As the pairing occurs, the deoxyribose attaches to the phosphoric acid of the preceding nucleotide, thus forming the strand of the new spiral.

Note that the replication process results in exact copies of the DNA molecules for hereditary transmission. An extremely interesting implication of the Watson and Crick theory of DNA structure is that the "code" that is passed from parent to offspring consists of sequences of base pairs along the DNA molecule. In other words, different sequences of base pairs represent different genes (i.e., codes/messages).

Testing a Hypothesis About DNA Replication

According to Watson and Crick's theory, when DNA replicates, the two original strands separate, allowing two new strands of DNA to form. Figure 11.5 is a diagram of the hypothesized replication process, showing only the letters of the base pairs that stick out from the strands. Of course, in addition to the base pairs, each strand consists, in theory, of a backbone of alternate molecules of deoxyribose and phosphoric acid. As previously mentioned and as shown in the figure, the replication process presumably produces exact copies of the original strands. Note in particular that each new double-stranded DNA molecule consists of one old strand and one new strand. How can this theoretical idea be tested? In other words, can we in fact find evidence to support or reject the idea that the replication process results in double stranded DNA molecules that are half old and half new?

In 1958 Matthew Meselson and F. W. Stahl performed an experiment designed to test this aspect of Watson and Crick's theory. In preparation for their experiment, they grew a population of bacteria for 14 generations, during which the bacteria "ate" a special diet. The special diet contained molecules of ammonium chloride (NH_4Cl), in which all the nitrogen atoms were the heavy isotope N^{15} instead of the normal lighter isotope N^{14}. Because the bacteria presumably replicated their DNA molecules 14 times during this time, each time producing DNA strands containing N^{15} atoms, Meselson and Stahl assumed that after 14 generations nearly all of the nitrogen atoms in the bacteria's DNA contained the N^{15} isotope. Another population of bacteria was grown as well. These bacteria were treated in precisely the same way as the others except that they "ate" ammonium chloride molecules that contained nitrogen atoms with the normal N^{14} isotope. Thus their DNA molecules presumably contained only the N^{14} isotope.

Meselson and Stahl's experiment consisted of using a centrifuge to extract and isolate the DNA molecules from some of each type of bacteria. The DNA molecules that contained the heavy N^{15} and those with the lighter N^{14} isotope were then poured into a test tube that held a

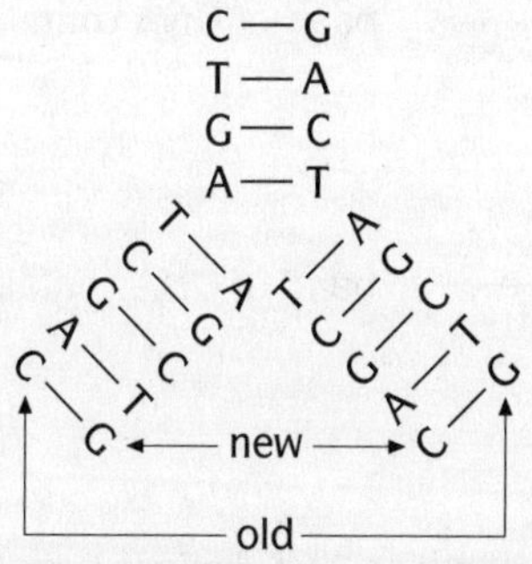

Figure 11.5 Diagram showing the way in which DNA replicates according to Watson and Crick's theory. The letters C, G, A, and T represent the bases cytosine, guanine, adenine, and thymine.

special liquid. The special liquid consisted of molecules of varying molecular weights so that when they were spun in the centrifuge they would form a gradient with the lighter molecules on the top and the heavier ones at the bottom. Meselson and Stahl then added the DNA molecules to the special liquid and spun them together in the centrifuge for about 48 hours. When they looked closely at the test tube, they discovered that two distinct bands appeared in the liquid (see Figure 11.6). Because it is reasonable to assume that heavier molecules had migrated more toward the bottom of the test tube, they concluded that the top band consisted of the N^{14} DNA molecules, while the bottom band consisted of the N^{15} DNA molecules.

Meselson and Stahl then collected some additional bacteria that contained the N^{15} DNA and placed them in a dish that contained ammonium chloride molecules with the normal N^{14} isotope. They let these bacteria stay in that dish just long enough to reproduce once (that is, replicate their DNA molecules one time). According to Watson and Crick's theory, this would mean that the DNA in this next generation of bacteria would consist of two strands, and one of those strands (the old one) should have N^{15} atoms and the other (the new one) should have N^{14} atoms in it.

Now, suppose that DNA extracted from these "hybrid" bacteria, DNA extracted from the N^{15} bacteria, and DNA extracted from the N^{14} bacteria were all poured into a test tube with the special liquid and

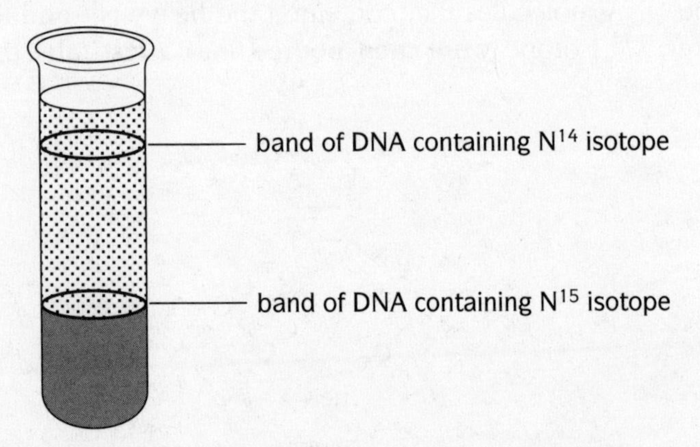

Figure 11.6 A test tube showing two distinct bands in the special liquid after being spun in a centrifuge. The top band presumably consists of lighter DNA molecules containing the N^{14} isotope. The bottom band presumably consists of heavier DNA molecules containing the N^{15} isotope.

spun in the centrifuge. According to the theory, what should happen? Meselson and Stahl's reasoning was as follows:

Hypothesis: *If*...the test tube contains N^{14} DNA, N^{15} DNA, and hybrid DNA that has one strand with N^{14} and one strand with N^{15}, and the mixture is spun in the centrifuge...

Prediction: *Then*... three distinct bands of DNA should show up, with the middle one appearing exactly halfway between the other two.

When Meselson and Stahl conducted the experiment and looked closely at the test tube, they discovered three bands in precisely the predicted pattern. The middle band showed up exactly halfway between the top and bottom bands (see Figure 11.7). They interpreted this to mean that the top band contained the N^{14} DNA, the bottom band contained the N^{15} DNA, and the middle band contained the two stranded DNA with half N^{14} strands and half N^{15} strands. Because this result was just what the theory led them to predict would happen, they concluded that this hypothesis of Watson and Crick's theory of DNA replication had been supported. In other words, evidence had been obtained that indicates that, when DNA replicates, it produces two strands of DNA in which one strand comes from the original DNA and the other strand is a new one. Of course, had the middle band not appeared halfway between the other two, this aspect of the theory would not have been supported.

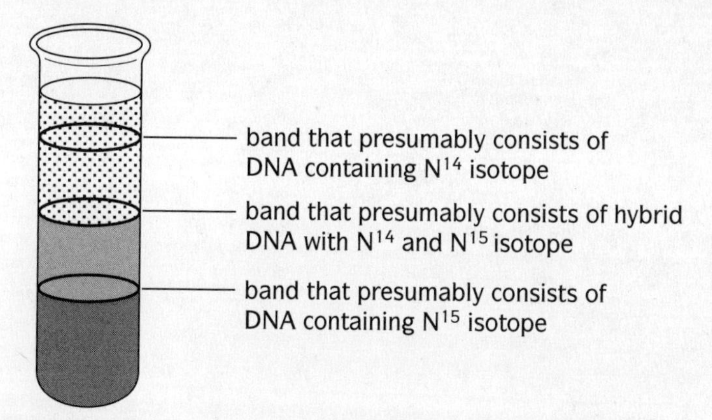

Figure 11.7 A test tube showing three bands in the liquid after being spun in a centrifuge. The finding that three bands appeared with the middle band half way between the other two was predicted by Watson and Crick's theory; therefore, this finding supported one aspect of their theory.

A THEORY OF GENE FUNCTION (PROTEIN SYNTHESIS)

According to the Watson/Crick theory, genes consist of sequences of base pairs along strands of DNA. But how do genes work to control the cell's activities?

The modern view is that they provide a template for the manufacture of complex molecules called proteins, and the proteins in turn control the cell's activities, primarily by controlling which reactions can occur and which cannot. Although there are many kinds of protein molecules, they all consist of sequences of smaller molecules called amino acids. There are only 20 different kinds of amino acid molecules, each consisting of a number of carbon, hydrogen, oxygen, and nitrogen atoms arranged in slightly different ways.

DNA molecules code for the manufacture of various kinds of proteins according to the following postulates.

POSTULATES OF A GENERAL THEORY OF GENE FUNCTION:

1. Instructions for making proteins are contained in the sequences of bases in the DNA molecule located in the cell nucleus.
2. A single amino acid is coded for by a sequence of three base pairs in a DNA molecule.
3. A strand of messenger ribonucleic acid (RNA) is formed in a nucleus complementary to its DNA strand.
4. The messenger RNA strand moves out of the nucleus, where ribosomes attach to it.
5. Transfer RNA's, one or more for each kind of amino acid, form in the nucleus and move out into the cytoplasm.
6. Amino acids in the cytoplasm become attached to their specific transfer RNA's.
7. A transfer RNA and its attached amino acid move to a ribosome that is attached to the messenger RNA, and the transfer RNA becomes attached to the messenger RNA at a position specified by the messenger RNA code, which consists of a sequence of three organic bases.
8. While in this position, the amino acid is joined to the preceding amino acid.
9. The ribosome moves along the messenger RNA, and the transfer RNA's bring in the different amino acids, so the manufacture of the protein is a sequential, linear process.

The process of manufacturing proteins according to this theory is depicted in Figure 11.8.

Testing Postulates of the Gene Action Function

How can the theory be tested? If we were small enough to crawl into a cell, we could simply sit back and observe the comings and goings of the various molecules and see if things happen as just described. The problem of course, is that we cannot crawl into cells, nor is the available technology good enough to allow us to film and magnify the molecular events as they take place. Nevertheless, the available technology is good enough to provide us with indirect ways of testing some of the theory's postulates. Let's see how this can be done.

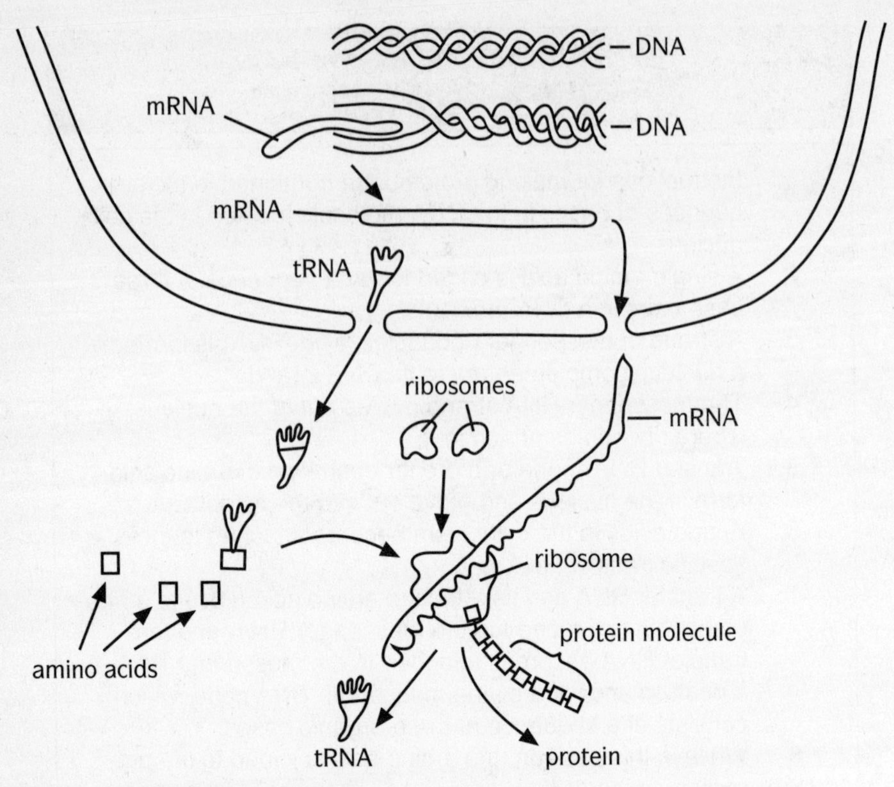

Figure 11.8 The major events in the synthesis of protein molecules as directed by the nuclear DNA in eukaryotic cells. mRNA = messenger RNA; tRNA = transfer RNA; rRNA = ribosomal RNA.

According to postulate 3 of the theory, genes control the manufacture of protein molecules in several steps. The first step in producing a specific protein presumably involves the production of a strand of messenger RNA (mRNA) complementary to a DNA strand located in the cell nucleus. The mRNA is supposed to be a single strand that has bases sticking out along the strand, just as is the case for the DNA molecule. The only significant difference between the two types of molecules is that the mRNA contains a base called uracil instead of the base thymine found in DNA. This fact plus the fact that uracil molecules can be made radioactive and that this radioactivity will show up in photographs as dark spots (the radioactive molecules presumably give off electrons that darken the film), provides an indirect way to test postulate 3 as follows:

Hypothesis: *If*...strands of mRNA are manufactured in the cell nucleus complementary to their DNA strands...

and...cells are grown for a period of time with "food" that contains radioactive uracil molecules...

Prediction: *Then*...the radioactive uracil molecules should make their way into the cell nucleus and gather near the DNA containing chromosomes. Thus, a photograph of the chromosomes should reveal many dark spots.

Photographic evidence to test this hypothesis and prediction was obtained by C. Pelling in Germany in the 1960s. Pelling's photograph showed chromosomes from the nucleus of insect larva cells. The predicted dark spots showed up in clear bands along the chromosomes, providing impressive evidence that supports the hypothesis that mRNA molecules are indeed being manufactured at these locations. It appears as though specific genes are indeed located in bands on chromosomes.

Using a similar technique, additional photographic evidence to support another postulate of the theory was gathered in 1969 by O. Miller and B. Beatty in the United States. Postulate 4 of the theory states that the mRNA moves out of the cell nucleus into the cytoplasm to begin directing the process of combining amino acid molecules into larger protein molecules. To test this postulate Miller and Beatty reasoned as follows:

Hypothesis: *If*...mRNA actually moves from the cell nucleus out into the cytoplasm...

and...cells with radioactive mRNA are photographed when the mRNA is being manufactured and again later after the source of radioactive "food" has been removed...

Prediction: *then*...initial photographs should show dark spots in the cell's nucleus, but later photographs should show the dark spots out in the cytoplasm. (Presumably the radioactive mRNA has moved from the nucleus to the cytoplasm.)

Again, just as was predicted by the theory, Miller and Beatty's photographs show dark spots in the nucleus in the initial photograph and in the cytoplasm in the later photographs. Thus, even though we are not able to see the hypothesized molecules of mRNA, we have obtained evidence that supports the part of the theory that claims that they are manufactured in the cell nucleus by the cell's DNA and that they then travel into the cell's cytoplasm. Although additional experiments have provided support for the theory's other postulates, many of the details of protein synthesis are still being investigated.

QUESTIONS

1. For a quick source of energy, which do you think would be better, a candy bar or a green salad? Explain.

___

___

___

___

2. Suppose a 10k marathon was to be run at sea level and another 10k marathon was to be run in the mile-high city of Denver, Colorado. Which location would you expect to produce the better times for the race? Explain.

___

___

___

___

3. In what ways is a burning log in a fireplace similar to cellular respiration? In what ways are the processes different?

___

___

___

___

4. Figure 11.1 shows in a general way the sequence of reactions that presumably take place during respiration (for example, carbohydrates → sugars → pyruvic acid → acetic acid → citric acid → water + carbon dioxide). Assuming that you did not know the sequence in which these reactions occurred but that you were able to detect each of these types of molecules, how could you research to determine the correct sequence?

5. Suppose you place a tiny fan with blades painted black on one side and silver on the other into a jar in which all the air has been removed. When light shines on the black side of the blades, will the blades rotate? Explain.

6. According to the modern theory of photosynthesis, photosynthesis occurs inside cell organelles called chloroplasts. Suppose this idea were challenged by a skeptic who claimed that photosynthesis really takes place in the cytoplasm but outside the chloroplasts. How could you experiment to test these alternative ideas? What are the predicted results of your experiment according to each idea?

7. Suppose that Meselson and Stahl allowed the N^{15} containing bacteria to reproduce for two generations in the container with the N^{14} containing food supply. Suppose further that a mixture of the

DNA from these second-generation bacteria were poured into a test tube that contained the special liquid and spun in a centrifuge. How many bands of DNA would you expect to find and where would it/they be located relative to where N^{14} and N^{15} bands appeared before?

___

___

___

___

8. Explain what the lines on the graph below mean in terms of the Meselson-Stahl experiment? What would the line look like after four bacterial generations? (The graph appears at end of chapter on page 269.)

___

___

___

___

9. If it is true that genes code only for the manufacture of protein molecules, where might the cell obtain the other basic types of molecules needed for survival, such as nucleic acids, carbohydrates, and lipids?

___

___

___

___

10. What forces can you think of that might be responsible for the movement of mRNA out of the cell nucleus?

___

___

___

___

11. Assuming that there is no tiny traffic cop inside cells, how do you suppose the molecules involved in the synthesis of proteins manage to end up in the right places at the right times?

____

____

____

____

12. Evidence has been obtained to support the hypothesis that a gene is a segment of the DNA molecule that is capable of coding for the manufacture of protein molecules. How long would a segment of DNA have to be to code for a protein? (Hint: Assume that DNA consists of various sequences of just four different bases—A, T, G, C. Assume also that protein molecules consist of chains of amino acid molecules and that there are 20 different kinds of amino acids. Therefore, protein molecules differ from each other in the number and sequence of their amino acids. This implies that the DNA code would have to be larger than a single base—a single letter. If the code consisted of a single letter (there are only four letters) then only four different amino acids could be coded for. Since there are 20 amino acids, the code must contain more than one letter (more than one base). Would a two-letter code work? How many different amino acids could a two letter code specify? How about a three-letter code, and so forth?).

____

____

____

____

TERMS TO KNOW

adenosine diphosphate	Krebs cycle
adenosine triphosphate	light-dependent reactions
Calvin cycle	light-independent reactions
cellular respiration	photon (photo-, light)
chloroplast	photosynthesis
electron transport chain	radioactive
fermentation	thylakoids

SAMPLE EXAM

Circle or enter the letter of the best answer.

1. Some algae cells have only one mitochondrion. What can you infer about the energy needs of these cells?

    a. They are very low.

    b. They are very high.

    c. They fluctuate, depending on temperature.

    d. More information is needed.

2. Investigators have cultured cells in a medium that contains radioactive molecules. Within 20 minutes, the material has been taken into the cell. Within 60 minutes, the molecules are found throughout the endoplasmic reticulum. This evidence supports the hypothesis that the endoplasmic reticulum

    a. deals only with radioactive materials.

    b. is a transport system within the cell.

    c. has an unknown function.

    d. is active in breaking down harmful materials.

The next five items (3–7) refer to the following experimental design. One gram of dry baker's yeast was added to six test tubes of equal size and mixed with 30 ml distilled water or 30 ml of a 5% sucrose solution. The test tubes were then placed under different conditions of light and temperature as shown in Table 11.1. BTB was added to each solution. The color of the solution at the start of the experiment is shown. Note that CO_2 was bubbled into test tubes 1 and 3 to produce a green BTB solution. BTB was added to each test tube at the end of the experiment and the color of the solution was noted.

3. Solution type, amount of light, temperature, and color of BTB solution at that start of the experiment are
   a. variables held constant.
   b. independent variables.
   c. dependent variables.
   d. values of the experiment.
   e. possible results.

4. The color of the BTB solution at the end of the experiment is a(n)
   a. variable held constant.
   b. independent variable.
   c. dependent variable.
   d. value of the experiment.
   e. possible cause.

TABLE 11.1 EXPERIMENTAL CONDITIONS

TEST TUBE NO.	SOLUTION TYPE	AMOUNT OF LIGHT	TEMPERATURE (°C)	BTB COLOR AT START
1	distilled water	light	10°	green*
2	5% sucrose	dark	23°	blue
3	distilled water	dark	10°	green*
4	5% sucrose	dark	10°	blue
5	5% sucrose	light	10°	blue
6	distilled water	dark	23°	blue

*Color change from blue → green → yellow indicates the production of CO_2.

5. Which test tubes can be compared to test the effect of amount of light?

   a. 2 and 4

   b. 2 and 6

   c. 1 and 3

   d. 4 and 5

   e. 1 and 6

6. Assuming that the hypothesis that yeast are like animals and take in O_2 and expel CO_2 is true, what is the predicted (expected) color of the BTB solution at the end of the experiment in test tube number 2?

   a. blue

   b. green

   c. blue or green

   d. dark blue

   e. green or yellow

7. At the end of the experiment, the solution in test tube 5 turned greenish/yellow while the solution in test tube 6 remained blue. What caused the solutions to differ in color?

   a. amount of light

   b. temperature

   c. the solution type

   d. There is no way to tell.

   e. the yeast

The next four questions (8–11) are based on the following information and key.

A virus that attacks bacterial cells is grown with radioactive phosphorous (P*) attached to its DNA and radioactive sulphur (S*) attached to its outer protein coat. The virus is then placed in a bacterial culture. Below are possible results of this experiment:

I. P* is found inside bacterial cells.

II. S* is found inside bacterial cells.

III. P* is found outside bacterial cells.

IV. S* is found outside bacterial cells.

8. Which result would indicate that DNA from the virus had entered the bacterial cell?

   a. I

   b. II

   c. III

   d. IV

9. From result III one could tentatively conclude that

   a. the virus protein had mutated.

   b. the virus protein had not entered the bacteria.

   c. the virus DNA had not entered the bacteria.

   d. the virus does not contain DNA.

10. Which result would indicate that material from the outer coat of the virus had entered the bacterial cell?

    a. I

    b. II

    c. III

    d. IV

11. Which results would indicate none of the viruses had entered the bacterial cell?

    a. I and II

    b. II and III

    c. II and IV

    d. III and IV

The next four questions (12–15) are based on Table 11.2 and the key. Use the key to classify the statements.

KEY: a. The data provide evidence for this statement.

b. The data provide evidence against this statement.

c. The data provide no evidence either for or against this statement.

TABLE 11.2 RELATIVE AMOUNT OF NUCLEOTIDES IN DNA

SOURCE	ADENINE	THYMINE	GUANINE	CYTOSINE
Calf	1.13	1.11	0.86	0.85
Rat	1.15	1.14	0.86	0.82
Moth	0.84	0.80	1.22	1.33
Virus	1.17	1.12	0.90	0.81
Sperm (of Rat)	1.15	1.09	0.89	0.83

12. The DNA molecule is composed of nucleic acids, deoxyribose and phosphoric acid. ____________

13. There is specific pairing of nucleotides in the DNA molecule. ____________

14. Diameter constancy of the DNA molecule comes about by purine-pyrimidine pairings. ____________

15. The ratio of adenine to guanine is fairly constant for all species. ____________

16. There are at least two major processes in photosynthesis, a reaction controlled by light and a reaction controlled by enzymes. Evidence for this would be that

    a. at maximum light intensity an increase in temperature will increase photosynthesis.

    b. photosynthesis will not occur in the dark.

    c. photosynthesis must occur within a certain temperature range.

    d. photosynthesis will not occur in the absence of at least a certain amount of CO_2.

17. If a small fish was placed in a beaker of water that contained Bromthymol blue, within one hour the water would become yellow. The change to yellow would occur because of

    a. a decrease of CO_2.
    b. a decrease of O_2.
    c. an increase of CO_2.
    d. an increase of O_2.

18. If a fish in a beaker of water that contained Bromthymol blue and some *Elodea* were placed in a lighted area, there would be no change of color in the water because the photosynthetic activity of Elodea counteracts the

    a. decrease of CO_2.
    b. decrease of O_2.
    c. increase of O_2.
    d. increase of CO_2.

19. Which of the following is a predicted outcome of an experiment designed to test the hypothesis that chlorophyll molecules are necessary for photosynthesis?

    a. Plants in the dark should not be able to conduct photosynthesis.
    b. Albino plants should not produce starch once the energy in their cotyledons is depleted.
    c. Lack of CO_2 will limit the rate of photosynthesis.
    d. Albino plants should not be able to germinate.
    e. Only plants with chlorophyll will be able to conduct photosynthesis.

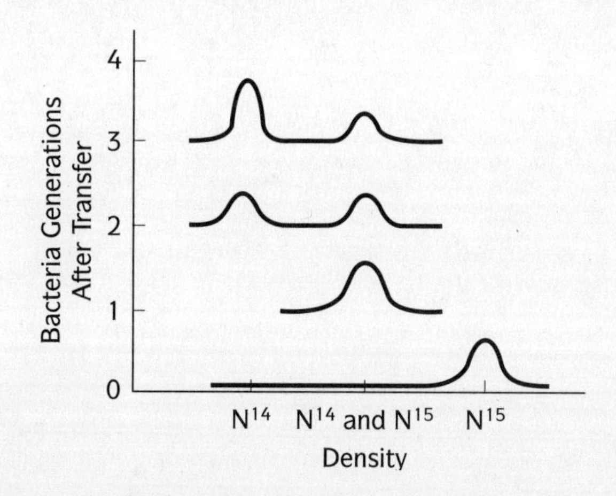

CHAPTER 12

ECOLOGICAL LEVEL THEORIES

GETTING FOCUSED

- *Identify a theory of population growth and crash.*
- *Learn about theories of population regulation.*
- *Learn about a theory of ecosystem dynamics.*
- *Learn about a theory of succession.*
- *Evaluate how ecological level hypotheses are tested.*

A THEORY OF POPULATION GROWTH AND CRASH

Under ideal conditions populations have an inherent ability to produce offspring in tremendous numbers. In other words, they generally have a high **biotic potential,** defined as the potential of a population to reproduce. For example, a population of horseflies has a high biotic potential. One pair of horseflies starting to breed in April could have 191,010,000,000,000,000,000 descendants by August if all of their eggs hatched and if all the young survived to reproduce. How can so many offspring be produced in such a short time? The answer is reflected by Figure 12.1. The figure reveals that the rate of population increase accelerates. In other words, 2 adults could produce 4; those 4 could produce 8; those 8 could produce 16; then 32, 64, 128, 256, 512 and so on. The rate of increase accelerates from generation to generation so the increases soon become astronomically large.

However, because of lack of food, space, water, disease, and the like, real populations do not increase in size indefinitely. Such factors, called **limiting factors,** act to produce a growth curve that looks more like that for the single-cell yeast population shown in Figure 12.2. The yeast population growth curve shows an S-shaped growth pattern, as initial increase is slow but accelerates rapidly until cell deaths start to outnumber cell births when limiting factors begin to assert themselves. The population stops growing when a balance is struck between the

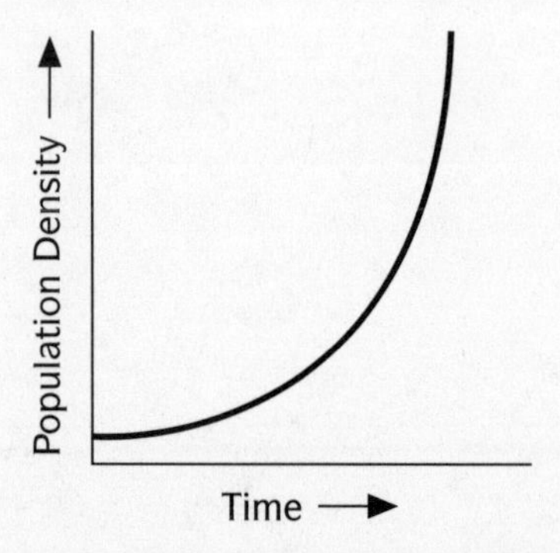

Figure 12.1 The exponential curve. The rate of increase accelerates until, in theory, the population density increases at an infinitely high rate.

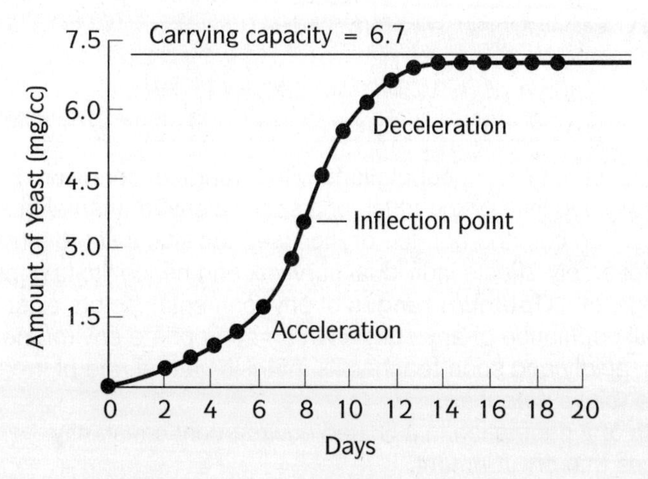

Figure 12.2 The growth curve of a population of yeast cells. The curve is often referred to as an S-shaped curve because of its increasing rate of growth at low densities, its inflection point, and where the rate of change shifts from increase to decrease. The decrease continues until the population size reaches the environment's carrying capacity. At this point the population size may stay near the carrying capacity or may decrease sharply.

population's biotic potential and the environment's ability to support the population. When the number of births equals the number of deaths, the population has reached what is referred to as the environment's **carrying capacity.** Here the term **environment** refers to the organism's surroundings, which consist of both living (biotic) and nonliving (abiotic) factors in its **habitat** (i.e., the place where it lives).

In environments where necessary raw materials and energy are continually resupplied, populations may remain near the carrying capacity indefinitely. Otherwise the population will decline rapidly as resources are depleted. The population may completely crash, as would be the case of a yeast population growing in a closed bottle.

POSTULATES OF A GENERAL THEORY OF POPULATION GROWTH

1. Environments present populations with a range of conditions.
2. Generally, too little or too much of a specific environmental factor (e.g., too cold/too hot, too dry/too wet, too little salt/too much salt) adversely affects individual survival, and hence limits population growth. **Optimum ranges** of environmental factors exist.
3. A small population of any species in an appropriate environment grows rapidly and soon reaches its intrinsic natural rate of increase (biotic potential).
4. Growth of a population in a limited environment constantly changes that environment.
5. Growth continues at the intrinsic rate until competition for resources and/or some effect of the environment, such as predation, reduces the rate.
6. Population growth stops when birth rate equals death rate.
7. If environments adversely change and/or necessary raw materials are not resupplied, populations will crash.

Testing a Postulate of the Theory

The second postulate states that too much or too little of a specific substance adversely affects individual survival and therefore limits population growth. Although this idea may seem intuitively correct, like any postulate it should not be accepted without testing. How can it be tested?

Table 12.1 reports data gathered by Aumann and Emlen, showing the relationship between the amount of salt found in soil samples and the density of a meadow vole population in the woodlands of the eastern United States. Notice that the soil with the lowest salt concentration has the lowest number of meadow voles living on it. Also, the soil with the highest salt concentration has the highest number. In general, the more salt the soil has in it, the more voles are found living on it. In other words, there is a correlation between the salt concentration and the density of the vole population. Could the lack of salt in the soil be limiting the size of the meadow vole population?

TABLE 12.1 RELATIONSHIP BETWEEN SALT LEVEL IN SOIL AND POPULATION DENSITY OF MEADOW VOLES

SALT LEVEL	NUMBER OF VOLES PER ACRE
Low	230
Medium	400
High	1000+

Data from G. D. Aumann and J. T. Emlen (1965) *Nature*, 208: 198–199.

To find out, Aumann and Emlen designed and carried out a series of experiments in which voles were actually bred under varying conditions of salt concentration. During the experiment some of the voles were given distilled drinking water that contained 0.5 percent salt solution. Other voles were bred under identical conditions except that they were given distilled water with no salt. Aumann and Emlen reasoned as follows:

Hypothesis: *If*...too little salt in the voles' diet does limit their ability to reproduce and in turn their population size...

Prediction: *then*...the voles without salt in their drinking water should produce fewer offspring than those with salt in their drinking water.

The results of the experiment were just as predicted. The voles with salt in their drinking water reproduced about 50% more offspring than the others during the 4 to 5 month test period. The result, therefore, supports the hypothesis that the lack of salt is a factor limiting vole populations. Just what it is about salt that is needed for high reproduction is not clear, but Aumann and Emlen hypothesized that an adequate salt concentration is needed for the proper functioning of the voles' adrenal glands.

THEORIES OF POPULATION REGULATION

Figure 12.3 shows one of the classic examples of predator-prey population fluctuations. The Hudson Bay Company of Canada contracted with trappers across Canada and the northern United States territories to buy their pelts. The company then shipped the pelts to England, where they were marketed. Since the early 1800s, the Hudson Bay Company has kept accurate records of all the pelts their trappers have taken each year. These data can now be used to indicate population sizes for the fur-bearing animals taken.

From the graph one can see the lynx population peaks on an average of every 9.6 years. During these fluctuations, the hare population seems to exhibit the same pattern, only the extremes are wider and the peaks seem to occur slightly before the lynx peaks, though these same data could be interpreted as following the lynx peaks. In this type of predator-prey interaction, the general explanation for the peaks and valleys goes something like this: As the hare/prey population increases, the lynx/predators have unlimited food, and their numbers increase tremendously. As the population of the prey continues to increase (usually exponentially), it soon approaches or reaches the carrying capacity of the environment, food supply usually being the limiting factor. With the food supply being depleted, competition among members of the prey population increases and the prey population crashes, followed shortly by the crash of the predator population for the same rea-

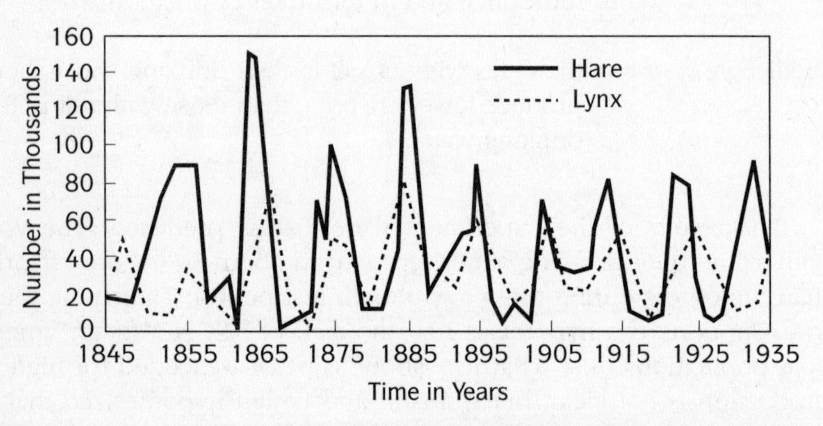

Figure 12.3 Fluctuations of lynx and hare populations in Canada from 1845 to 1935 as estimated by the number of animals captured by fur trappers. After D.A. MacLulich, University of Toronto Studies, Biology Series No. 43, 1937.

son. Such competition between members of the same species is called **intraspecific competition** as apposed to competition for limited resources between members of different species, which is called **interspecific competition.**

The data seem to indicate that intraspecific competition is very important in limiting population size. Interestingly, studies in areas where the snowshoe hare does not have any predators seem to indicate that the hare population undergoes fluctuations of a similar period on its own. Certainly intraspecific competition for food and shelter would increase in the hare population as its numbers increase. The following example of the Kaibab deer population in northern Arizona makes this point as well.

In 1906 President Theodore Roosevelt created the Grand Canyon National Game Preserve to help develop a large deer population on the Kaibab Plateau (North Rim area of the Grand Canyon). Deer hunting was thus prohibited. There were domestic grazing animals, mostly sheep and some cattle, but their numbers were decreasing at this time. Partially as a part of the game preserve plan, and partially in response to pressure from ranchers, hunters, and the general public, a predator removal program was started in 1906 to protect the deer herd. Over a period of 25 years (1906–1931) 6,254 predators were removed by hunting and trapping, including mountain lions, wolves, coyotes, and bobcats. During the years 1931–1939 public hunting of lions, coyotes, and bobcats removed 2,843 more predators. By 1931 there were no more wolves left in the area and other predator numbers had been greatly reduced.

The plan to create an ideal habitat for deer seemed to be a success. During this time the deer population increased from an estimated 3,000 to 4,000 in 1906 to an estimated 25,000 to 100,000 in 1924. At this point the carrying capacity of the environment had been exceeded. With no place to migrate to and the natural forage of the area all but wiped out, over the next 14 years intraspecific competition among the deer population increased tremendously and several thousands of deer died on the Kaibab, only a small fraction of which were killed by hunting (again permitted since 1924). The others died mostly from starvation and malnutrition-related causes.

Because intraspecific competition does not account for all of the evidence regarding population change, the major postulates of three alternative theories are presented below as a summary.

POSTULATES OF NICHOLSON'S THEORY

1. Animal populations are normally in a state of balance and, though they fluctuate, they do so in a restricted manner.
2. This situation is brought about only by factors that depress the population at high densities and increase it at low densities.
3. Reproductive rate, mortality due to lack of food, predation, competition, disease, and such self-regulating behavior as territorial fighting are the chief "density-dependent" factors (i.e., factors whose importance depends upon the density of the population).

POSTULATES OF ANDREWARTHA AND BIRCH'S THEORY

1. Most animal populations fluctuate irregularly through factors that act independently of density, notably those linked with climate.
2. Many populations become extinct and many populations persist through chance.
3. The ability of animals to migrate and thus to recolonize is of special importance.

POSTULATES OF WYNNE-EDWARDS'S THEORY

1. Animal populations are regulated by density-dependent factors.
2. Food shortage is the ultimate limiting factor.
3. Animals normally regulate their own density far below the limit set by food through dispersive behavior and restraints on reproduction.

Testing Postulates of the Theories

The Nicholson theory and the Wynne-Edwards's theory emphasize the importance of "density-dependent" factors in regulating animal populations. Density-dependent factors are those that vary depending upon the density of the population in question. For example, both lack of food and lack of space are density-dependent factors since their influence increases as population size increases. On the other hand, Andrewartha and Birch's theory emphasizes the importance of density-independent factors. For example, a night of subzero temperatures may kill all the members of a cricket population regardless of its size.

How can one determine whether density-dependent or density-independent factors are regulating the size of a specific population? An interesting experiment to discover the factors involved in regulating the population size of two species of barnacles living on rocks of the Scottish coast was conducted by the American ecologist Joseph Connell.

Barnacles are small volcano-shaped animals that cement themselves to rocks along the seashore. If you went to the area between the high- and low-tide marks in Scotland, you would find two kinds of barnacles living there. Up high on the rocks, you would find a population of rather small barnacles called *Chthamalus*. On the middle and lower rocks, you would find a population of slightly larger barnacles called *Balanus*. Why don't the smaller barnacles live down lower? And why don't the larger barnacles live up higher?

By carefully observing the barnacles over a long period of time, Connell found out that the larger barnacles actually pried the smaller ones off the rocks or simply grew over them when they happened to start growing in the same area. This is an example of a very direct competition for the limiting resource of available space. But why don't the larger barnacles take over in the upper area as well? Through experiments Connell found out that space is not the only factor limiting the barnacle populations.

Connell selected two similar locations where the two kinds of barnacles were living. In both locations, he scraped off all the barnacles from an area just above and below the boundary line where the two kinds of barnacles met. In the first location, he continuously kept the area free of small barnacles. In the second location, he continuously kept the area free of the large barnacles (see Figure 12.4 on page 280). The reasoning guiding Connell's experiment was as follows:

Hypothesis 1: *If*...density-independent factors regulate the distribution of these barnacles (i.e., too dry or too wet)...

Prediction 1: *then*...the large barnacles should not grow above the boundary line in location 1 (because it is too dry), and the small barnacles should not grow below the boundary line in location 2 (because it is too wet).

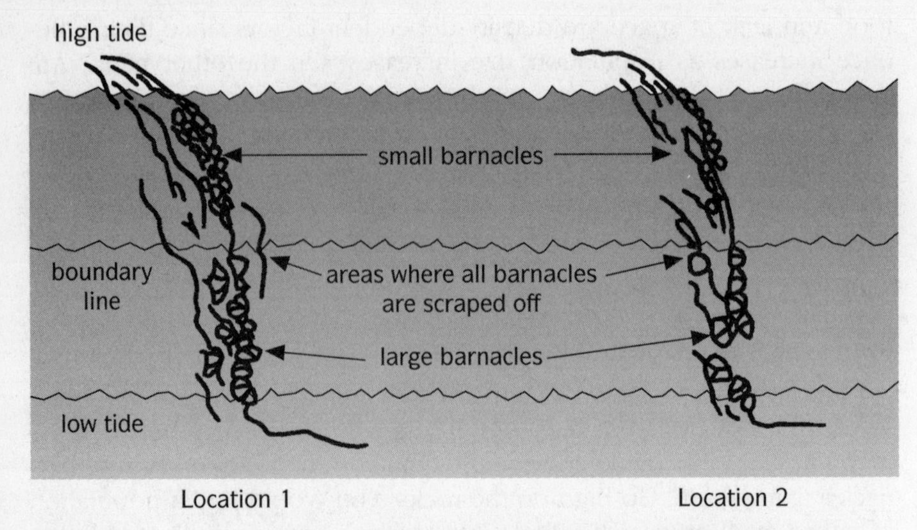

Figure 12.4 Connell scraped all the barnacles from near the boundary line between the two locations. In location 1, he continuously kept the area free of small barnacles. In location 2, he continuously kept the area free of large barnacles.

On the other hand:

Hypothesis 2: *If*...density-dependent factors regulate their distribution (that is, competition for space between the two kinds of barnacles)...

Prediction 2: *then*...the large barnacles should grow above the boundary line in location 1 (because competition for space with the small barnacles has been eliminated) and the small barnacles should grow below the boundary line in location 2 (because competition for space with the large barnacles has been eliminated).

The results of this experiment were that the large barnacles in location 1 were not able to grow above the boundary line as predicted by Hypothesis 1. This result suggests that the upper range of their distribution is controlled by lack of moisture, supporting the density-independent hypothesis. However, in location 2 the small barnacles were able to grow below the boundary line as predicted by Hypothesis 2.

This result suggests that the lower range of their distribution is controlled by competition with the large barnacles, thus providing support for the density-dependent hypothesis. In other words, Connell's experiment shows that, at least in the case of these two kinds of barnacles, both density-independent and density-dependent factors regulate animal populations.

A THEORY OF ECOSYSTEM DYNAMICS

Populations of interacting organisms living together in a particular area are referred to as a *biological community.* The populations of a biological community depend on each other directly or indirectly for survival. Survival also depends upon proper amounts of specific abiotic (nonliving) factors within the environment, such as light, heat, and various inorganic substances, such as oxygen, water, food, carbon dioxide, and nitrogen. Therefore, a biological community and its abiotic environment constitute a more complex system of interactions referred to as an ecological system or ecosystem. The populations will vary from one ecosystem to another, depending upon the abiotic environment. For example, an abundant supply of water may produce a forest, while lack of abundant water may produce a grassland. Although the specific sets of populations will vary from one ecosystem to another, the basic pattern of population interaction is the same in all ecosystems. That pattern involves the flow of energy into, through, and out of biological communities and the cyclic exchange of specific inorganic molecules between the biological community and its abiotic environment.

Both energy (in the form of sunlight) and specific inorganic molecules enter the community through green plants. Green plants, through the process of photosynthesis, are able to utilize light energy to synthesize complex organic molecules from the inorganic molecules absorbed from the environment. These complex organic molecules then serve as a food/energy source for the plants and indirectly for the animals of the community. Therefore, the plants are collectively referred to as **producers** and the animals that eat the producers are called first-order or primary **consumers** (the herbivores).

Animals that obtain food/energy by eating first-order consumers are called second-order or secondary consumers (the **carnivores**). Animals that eat second-order consumers are called third-order consumers, and so on. Some animals eat both plants and other animals and fit into the community at more than one feeding or **trophic level** (the **omnivores**). Still other animals obtain food/energy by eating dead organisms. These are the **scavengers.** An organism's role or "job" in the community is referred to as its **ecological niche.**

One final group of organisms exists in communities. This group, the **decomposers,** also obtains its food/energy from dead organisms. Decomposers, such as mold and bacteria, do not have mouths. Instead they absorb the molecules of dead organisms directly through their cell walls. The excretion of waste molecules by animals and the breakdown of organic molecules by decomposers are two essential processes that replenish the supply of inorganic molecules in the environment so that plants can absorb them and start the process all over again.

A specific sequence of feeding relationships that takes place in a specific biological community, such as diatoms → krill → penguin → killer whale (for the arrows read "is eaten by") is called a **food chain.** Most biological communities have many food chains with interconnecting side branches. The total network of feeding interactions is called a **food web.** Figure 12.5 shows the food web of the biological community of the Antarctic waters. Note that none of the food chains in this web is longer than five links. Seldom are food chains longer than this primarily because more and more of the energy that is input at the bottom (in this case the diatoms are the producers) is utilized by individuals at each level up the chain. Said another way, killer whales are too scarce and too hard to catch for you, or anyone else, to make a living by eating them.

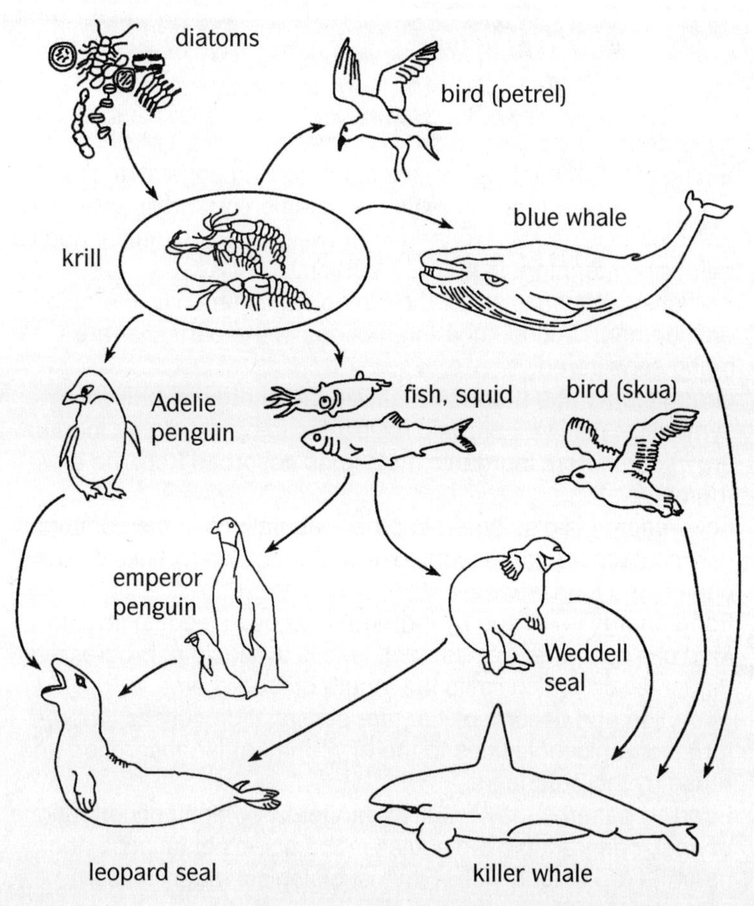

Figure 12.5 A food web in Antarctic waters composed of several food chains. Diatoms are the producers that directly or indirectly support the entire community.

Unlike energy, some substances may actually become more concentrated as they pass up food chains. If a poisonous molecule, such as the pesticide DDT, is sprayed on the surface of a pond, it will be absorbed by algae cells at relatively low concentrations. But because DDT is neither readily excreted nor broken down by organisms (i.e., it is nonbiodegradable), any animal, such as a mayfly, that eats lots of algae will consume a lot of DDT. If fish in turn eat a lot of mayflies, the DDT concentration in the fish will be even higher than it was in the mayflies. The end result is that animals at or near the top of food chains may accumulate deadly levels of DDT or any number of such toxic substances. The concentration of molecules by food chains in this way is called **biological magnification.**

POSTULATES OF THE THEORY OF ECOSYSTEM DYNAMICS

1. Biological communities consist of interacting populations in which energy enters and exits the community, and specific inorganic molecules are exchanged between the community and its abiotic environment in a cyclic fashion.
2. An ecosystem consists of the biological community, the abiotic environment, and all their interactions in the particular area being considered.
3. Green plants (the producers) utilize the energy of sunlight to synthesize complex organic molecules to be used as a food/energy source from inorganic molecules absorbed from the environment.
4. Food/energy is distributed to other populations in the community (the consumers and decomposers) through many links in food chains and food webs.
5. Food/energy is utilized by individuals at each feeding level in food chains; therefore, less and less is available at progressively higher levels, which limits the length of food chains.
6. Excretion and decomposition are essential processes that return inorganic molecules to the environment for absorption and reuse by the producers.
7. Feeding patterns may result in dangerously high concentrations of nonbiodegradable molecules in the bodies of animals at or near the top of food webs (that is, biological magnification).

Testing a Postulate of the Theory of Ecosystem Dynamics

Postulate four of the theory states that green plants act as producers within the ecosystem by utilizing the energy of sunlight to synthesize complex organic molecules from inorganic molecules absorbed from the environment. Part of this hypothesis can be tested as follows:

Hypothesis: *If*...growing plants absorb inorganic molecules from the environment...

and...growing plants are removed from the environment...

Prediction: *then*...the amounts of inorganic molecules in the environment should increase (through the continued action of decomposers on dead organisms).

Large-scale experiments to test this hypothesis and related postulates were begun in 1963 by a group of ecologists in a deciduous forest ecosystem in New Hampshire called the Hubbard Brook Experimental Forest. The study area consisted of a group of six forest-covered valleys, each with its own creek running down the middle.

The first task undertaken was to measure the inputs and outputs of water and other important inorganic molecules to and from the undisturbed forest. To do this, precipitation gauges were placed throughout the forest to measure the inputs of rain, snow, and dissolved inorganic molecules. Concrete channels were built across the creeks at the bottoms of the six valleys. Since the channels were built on bedrock, all the water that had not evaporated and all the inorganic molecules leaving the forest ecosystem had to pass through the channels and could be measured.

Data from the undisturbed forests revealed that the forest ecosystem was very efficient at retaining inorganic molecules. Inputs of nutrient molecules in precipitation were approximately equal to the nutrient outputs in the creeks, and both were small relative to the total amount of nutrients estimated to be present in the ecosystem.

During the winter of 1965–1966 all the trees and shrubs in one of the six valleys were cut down, leaving them where they fell. Regrowth was prevented by spraying with herbicides. The ecologists then resumed collecting data on the water and nutrient inputs and outputs from the deforested area. Dramatic differences became obvious almost from the outset. In comparison to the previous data and to the remaining undisturbed (control) forests nearby, water runoff in the deforested area increased by 40%. This represents the amount of water that would have left the undisturbed ecosystem by transpiration (water loss through leaves; see Chapter 8). Also, the loss of inorganic nutrients dissolved in the creek runoff was 6 to 8 times greater than in the undisturbed forests. The level of nitrate loss, for example, between the forests and deforested areas is shown in Figure 12.6 on page 286.

According to the theory of ecosystem dynamics, the cause of this increase in inorganic molecules is the action of decomposer organisms on the sudden abundance of dead organic matter. Those nutrients are presumably being returned to the soil and then collecting in runoff and being carried down streams. Therefore, in conclusion, the hypothesis that plants in the ecosystem absorb inorganic nutrients from the environment is supported. The hypothesis led to the prediction that the

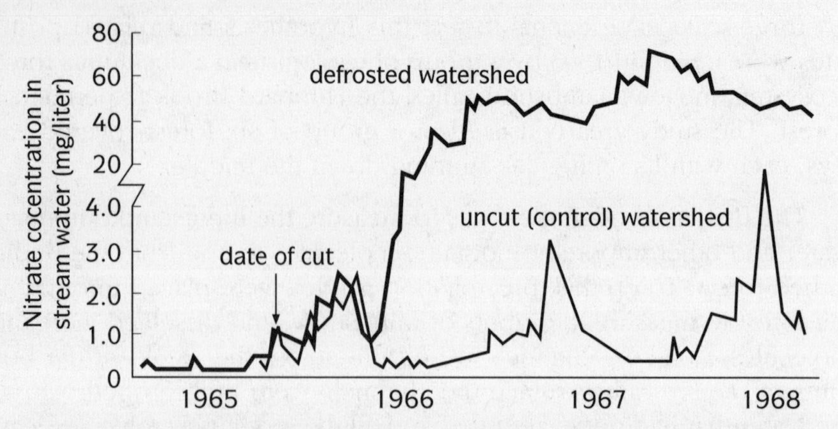

Figure 12.6 Nitrate concentration in the stream water in the deforested watershed and in the uncut (control) watershed at the Hubbard Brook Experimental Forest. The date of deforestation is indicated by the arrow. Note that a change in scale of the vertical axis was necessary to keep both lines on the graph. After Likens et al., "Effects of forest cutting and herbicide treatment on nutrient budgets in the Hubbard Brook watershed ecosystem," ***Ecological Monographs,*** **40, 23. Copyright 1970 by the Ecological Society of America.**

removal of the plants would cause an increase in the level of nutrients in the environment, and that is exactly what happened.

A THEORY OF SUCCESSION

Landslides, erosion, the eruption of a volcano, the retreat of a glacier, or the rise of new land from the sea all result in new areas that, for a while at least, are not occupied by living things. The colonization and subsequent series of changes in biological communities that inhabit these new areas is known as **succession.** Successional changes also take place on previously inhabited land that, for some reason or another, such as fire, or clearing, and plowing, has been laid bare. In both types of succession, sometimes referred to as primary and secondary succession respectively, the organisms themselves alter the environment and cause succession to progress.

Consider the primary succession of newly exposed bare rock. The first organisms to colonize the bare rock are usually lichens and mosses. They are the only organisms able to attach to the rock and survive without elaborate root systems. Through their respiration, the lichens and mosses create carbonic acid, which begins to break up the rocks. Cooling and thawing also contribute to the breakup. When the lichens and mosses die, their dead tissues begin to accumulate in the new-formed cracks in the rocks. Dust is able to settle and collect as well. Eventually, the buildup results in pockets of soil that can begin to support larger plants.

When windblown seeds arrive, they get caught in the cracks, and their growth contributes to the rock breakup. Eventually, the mosses and lichens are crowded out by the taller plants that grow over and shade them. Still larger plants and animals may now migrate in as soil continues to build up. The number of different species increases. All of this continues to alter the environment. As the soil continues to become enriched with the substances from the breakdown of more and more plants and animals, still other types of vegetation are able to move in and crowd out the earlier forms. Eventually, a relatively stable community called a **climax community** is reached. Members of the climax community continue to die, yet they are replaced largely by members of their own species rather than by new successional forms.

The pattern of community change during secondary succession is much the same. The succession of communities on abandoned farmland in the southeastern United States is shown in Figure 12.7. As shown, the fields are first colonized by crabgrasses, horseweeds, and asters. Within 3 to 15 years or so, broomsedge and small shrubs grow over and crowd out the smaller plants. In about 20 years, the shrubs are replaced by small pine seedlings that require almost full sunlight

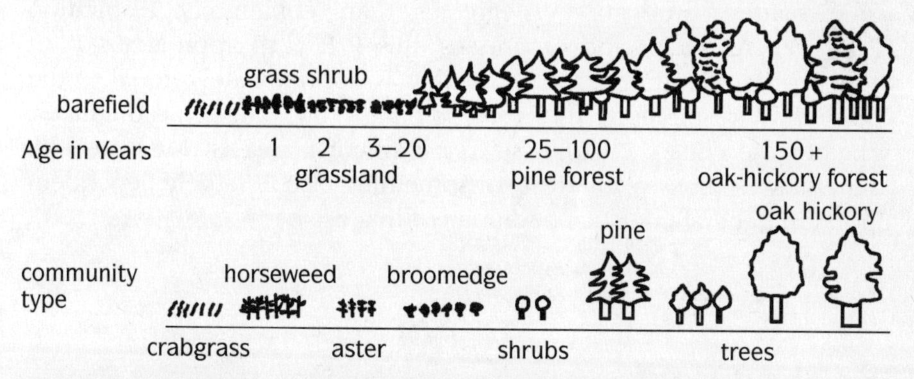

Figure 12.7 Secondary succession of abandoned crop land in the southeastern United States. After E. P. Odum, *Fundamentals of Ecology* (Philadelphia: Saunders, 1959), 263.

for growth. As the pines grow, they become denser and cast shadows on the ground and prohibit their own seedlings from developing. The seedlings of hardwood trees such as oak and hickory, however, can grow in the shade; thus they begin to grow, and after 90–100 years, they begin to crowd out the shorter pines. Not only can the oak and hickory seedlings grow in shade, but they require shade. This explains why they do not begin to grow in the area until after the pines have grown and shaded the ground.

When the oaks and hickory trees are tall, they not only are able to successfully compete with the pines for needed sunlight, but also through their deeper root system they are also able to obtain water when the pines are not. Thus, after 150–200, years the stages of succession result in an oak-hickory dominated climax forest. The forest community remains fairly stable because its dominant type of vegetation is able to perpetuate itself. Its seedlings are able to compete successfully in the environment. If nothing happens to drastically alter the climatic conditions of the area, or if fire or destruction by humans does not damage the community, it will remain essentially the same indefinitely.

Thus both primary and secondary succession generally produce communities with greater complexity. This complexity is reflected in longer food chains and more complex food webs involving more species of smaller populations, hence greater diversity and greater biomass (that is, total mass of living material). These structural changes during succession are accompanied by functional changes of the community. Early stages of succession show high rates of production of organic matter; production is greater than community respiration ($P/R > 1$). As succession progresses, this P/R ratio approaches 1. In other words, the amount of production is approximately equal to the community respiration; therefore, biomass no longer accumulates. Whether the climax community in a particular area is dominated by cacti, mosses, grasses, tall trees, or something else, is largely dependent upon the general climatic conditions of that area.

POSTULATES OF A GENERAL THEORY OF SUCCESSION

1. Species possessing the proper migratory and adaptive characteristics colonize a new or disturbed area and begin to modify the environment.
2. Succession occurs because species alter the environment in a way that makes it less favorable to themselves and more favorable for other species.
3. Competition among species and existing conditions determine which species persist and which decline or disappear.
4. Competition and stable conditions favor certain species that do not modify the environment and are able to reproduce in such a way that a stable climax community ultimately forms.
5. In general, as succession progresses, food web complexity, species diversity, biomass, and stability increase, and the ratio of community production to respiration decreases to approximately 1.
6. General climatic conditions and local environmental conditions determine the nature of the climax community (i.e., desert, grassland, forest) in any particular area.

Testing Postulates About the Causes of Succession

The third postulate of the theory of succession states that competition among species and existing conditions determine which species persist and which decline or disappear during succession. To test this postulate, ecologist W. P. Sousa studied the succession that takes place when rocks are scraped bare in the low intertidal zone of the Southern California coast. To mimic bare rock, Sousa placed some large concrete blocks in the water and carefully recorded the successional changes that took place on them from September 1974 to February 1977.

Figure 12.8 on page 290 shows some of the changes that took place. Initially, the cement blocks were colonized by the green algae *Ulva*. This green algae is a relatively short-lived and rapid growing plant that reproduces throughout the year. Accordingly, it was able to establish itself quickly on newly exposed surfaces. Within a year or so, however, the dominant plant was the more slow growing red algae

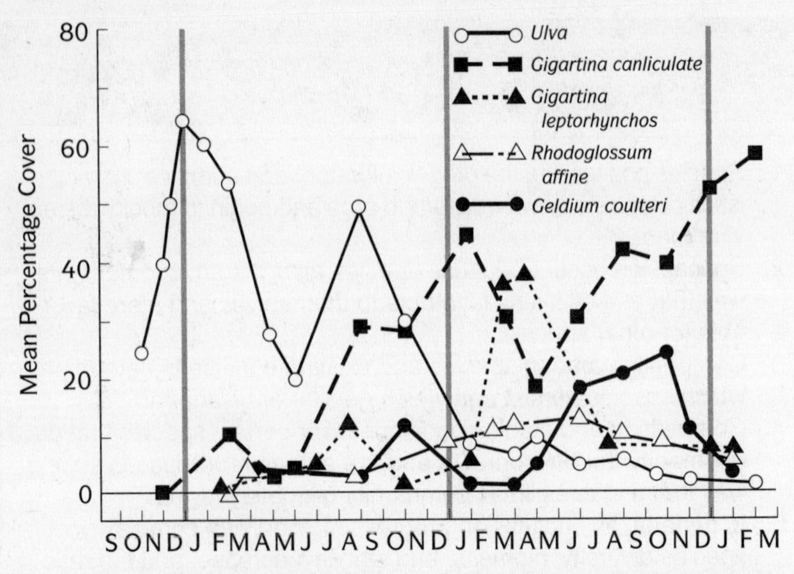

Figure 12.8 Mean percentage of cover of five algal species that colonize concrete blocks placed in the intertidal zone in September 1974. After Sousa, *Ecological Monographs,* 49, 1979, 227–254.

called *Gigartina canaliculata*. A second species of red algae called *Gigartina leptorhynchos* also increased in abundance and became dominant for a short time, while the abundance of the green algae steadily declined and eventually disappeared. What caused these successional changes? To find out, Sousa conducted a few experiments.

In one experiment, Sousa selected two identical cement blocks that were covered with the green algae. He then removed all the green algae from one (the experimental block) and left the other block alone (the control block). Over the next four months, he measured and recorded the number of red algae plants that grew on the blocks. The data, shown in Figure 12.9, indicate that the red algae quickly colonized the block that had the green algae removed, but did not gain much of a foothold on the block with the green algae. The green algae appear to be able to inhibit the colonization and growth of the red algae. How is it then that the red algae are ever able to take over?

As a possible answer to this question, Sousa advanced the hypothesis that the green algae are more in danger than the red algae from the intense sunlight and drying winds that exist when low tides occur during midday. Thus, when these conditions occur, the red algae will be able to survive and grow, while the green algae will dry out and die.

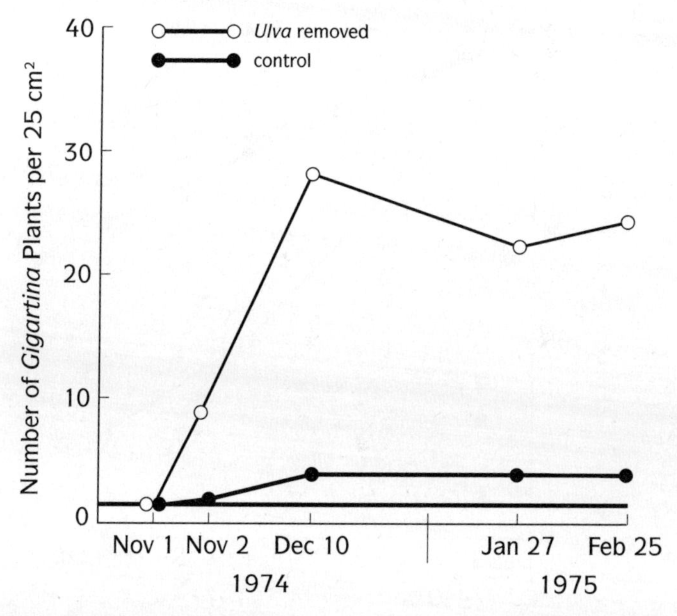

Figure 12.9 Effect of removing the early successional species *Ulva* on the colonization of *Gigartina* plants on four concrete blocks over a four-month period. After Sousa, *Ecological Monographs,* 49, 1979, 227–254.

To test his hypothesis, Sousa tagged and observed individual plants during a two-month period when low tides occurred during the early afternoon of sunny, dry days. Sousa reasoned as follows:

Hypothesis: *If*...succession occurs because sunny dry conditions favor the survival of red algae over green algae...

Prediction 1: *then*...the percentage of red algae plants that survive those conditions should be greater than the percentage of green algae plants that survive.

On the other hand,

Alternative Hypothesis: *If*...succession occurs for some other reason...

Alternative Prediction: *then*...the percentage of red and green algae plants should be the same.

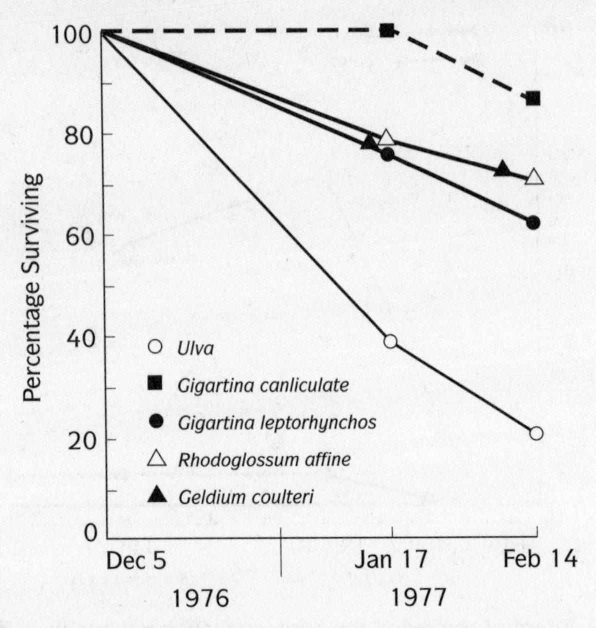

Figure 12.10 Survival curves for five species of algae over a two-month period when low tides occurred in the afternoon creating harsh physical conditions. Thirty plants of each species were initially tagged. After Sousa, *Ecological Monographs,* 49, 1979, 227– 254.

The experimental data are shown in Figure 12.10. As you can see, the percentage of surviving red algae plants (both species) was much greater than the percentage of surviving green algae plants. Because this result was predicted by Sousa's hypothesis, his hypothesis has been supported. Therefore, it appears that plant succession in this intertidal community occurs because the red algae are better able to compete when conditions become harsh. Sousa's specific hypothesis and the more general hypothesis that competition among species and existing conditions determine which species persist and which decline or disappear appear to be correct.

QUESTIONS

1. Suppose a population of bacteria that doubles every minute is placed in a jar. Suppose also that at exactly midnight the population has grown so that the jar is completely full of bacteria. At

what time was the jar only half full of bacteria? What is the implication of this hypothetical situation for understanding world human population growth?

2. Consider the human population in your area. Is the population still growing? If not, suggest several factors that may have stopped its growth. Which of these factors do you think were the most likely factors? Explain. If the population is still growing, suggest several factors that may limit future growth. Which of these factors do you think will be the most important limiting factors? Explain.

3. The two lines in Figure 12.11 on page 294 represent the adult population size of a predator and a prey species that live in one area over a period of 100 months. Which do you think is the predator and which do you think is the prey? Explain.

4. Figure 12.12 on page 295 shows graphs of population size over a 20-week period of two species of flour beetle and their larvae living in enclosed containers. Based upon the data depicted in

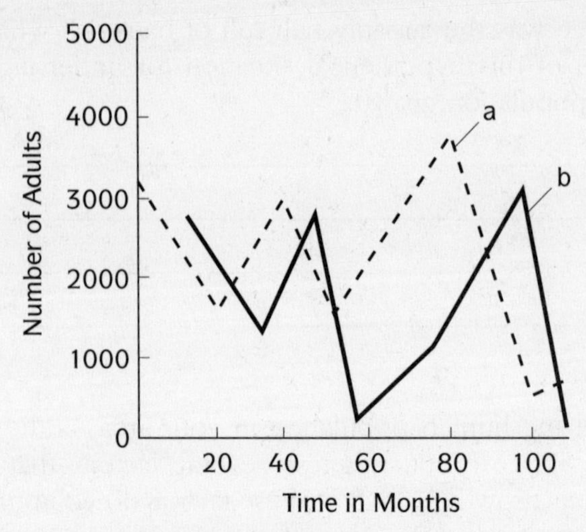

Figure 12.11 After J. J. W. Baker and G. Allen, *The Study of Biology,* 3rd ed. (Reading, MA: Addison-Wesley, 1977).

the graphs, evaluate each of the following statements as true or false. Please explain your choice in each case.

a. The experimental design allows one to test for the effect of temperature and the initial number of adults on population growth. Explain.

____

____

____

b. Comparing graph B with graph D allows one to conclude the adult flour beetle population grows better at 35°C than at 20°C. Explain.

____

____

____

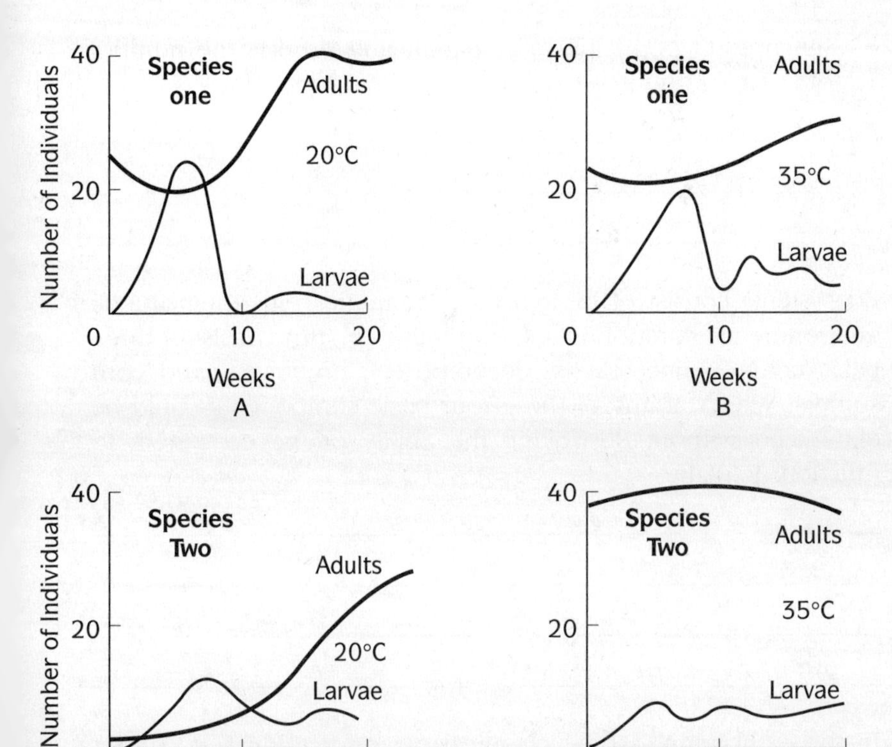

Figure 12.12 After J. J. W. Baker and G. Allen, *The Study of Biology,* 3rd ed. (Reading MA: Addison-Wesley, 1977).

c. Comparing graph A with graph B allows one to conclude that adult flour beetle populations grow better at 20°C than at 35°C. Explain.

____

____

____

d. The carrying capacity of this environment for adult beetles under optimal temperatures is probably about 40 individuals. Explain.

____

____

____

e. Initial adult population size significantly affects the number of larvae produced. Explain.

____

____

____

5. Ecosystems consist of biological communities plus their abiotic environmental components. Communities in turn consist of three basic components, that is, decomposers, producers, and consumers. Which, if any of these four basic components could be eliminated without destroying the ability of the ecosystem to function? Explain.

____

____

____

____

6. In the early spring, some chemical poison was dumped into a small lake. The poison killed all of the molds and bacteria and other types of decomposers. By the end of the summer members of the lake's bass population began to die. Rank the following from the most to least likely cause of the bass deaths. Explain your ranking.

a. The poison that killed the decomposers also killed the bass.

____

____

____

b. The bass had used the decomposers as a food source; since the food is no longer available, the bass starved.

____

____

____

c. Bass ate the poisoned decomposers and poisoned themselves.

____

____

____

d. The dead decomposers could no longer recycle nutrients, so the productivity of the pond dropped to near zero.

____

____

____

7. Sousa's experiment provided evidence in support of the hypothesis that sunny, dry environmental conditions favored the succession of red algae in place of green algae in the intertidal zone. But was this a "good" experiment in terms of controls? Can you think of one or more other reasons, other than the hot, dry conditions, why the red algae might have survived better than the green algae? How could these alternatives be tested?

____

____

____

____

8. In science, the answer to one question often leads to others. For example, now that we have evidence that red algae tolerate sunny, dry conditions better than green algae, we could ask why? That is, what is it about the red algae that make them more tolerant? Develop two alternative hypotheses to answer this question. How could you research to test the alternatives?

____

____

____

____

9. Regions of relatively stable climax communities that cover many miles of land are sometimes referred to as biomes. Consult reference material and list the world's major terrestrial biomes and their primary locations. Most likely your reference source indicates that the state of Arizona is part of the desert biome. But, in fact, all the world's major biomes except for tropical rain forest, can be found in Arizona. How can this be? In other words, what conditions must exist in Arizona to allow for the existence of such a wide range of climax communities? Explain in general terms how these conditions influence the plants that dominate the various communities.

____

____

____

____

TERMS TO KNOW

abiotic factors
biological magnification
biotic factors
biotic potential
carnivore
carrying capacity
climax community
consumers
decomposers
ecological niche
environment
food chain
food web
habitat
interspecific competition
intraspecific competition
limiting factor
omnivore
optimum range
producers
scavenger
succession
trophic level

SAMPLE EXAM

Circle or enter the letter of the best answer (e.g., a, b, c, d).

1. Several kinds of aquatic plants, algae, bacteria, one-celled protists, snails, and fish were placed in an aquarium. The aquarium was sealed. Each week the aquarium was observed and carefully weighed. The weight of the aquarium plus its contents would probably

   a. increase for a few weeks and then decrease.

   b. decrease for a few weeks and then increase.

   c. continually increase.

   d. continually decrease.

   e. not change.

Identify the following six statements (2–7) as to whether they are

a. a hypothesis

b. a prediction

c. an observation

d. a causal question

e. a conclusion

While walking around a lake one day, you find two dead fish lying about 10 feet apart on the shore. One of the fish you recognize as a bluegill, the other a bass. The bluegill is lying within 1 foot of the lake water on a moist, muddy area. The bass is resting on a dry, sandy area 6 feet from the water. Upon returning to the area two weeks later, you find the bluegill tissue is almost completely decomposed, whereas the bass is just beginning to decompose.

2. Several fly maggots can be seen crawling in and on both fish.

   ____

3. The bass might have tougher tissue.

____

4. The difference in decomposition may be due to the amount of available moisture.

____

5. If dead bluegill and bass are placed within one foot of the lake, they should decompose at a similar rate.

____

6. Since the fish were different types, they were exposed to different types of soil, and they were exposed to different amounts of water, we cannot say for sure why they are decomposing at different rates.

____

7. Why is the bluegill more decomposed than the bass?

____

8. Grass clippings from a recently mowed lawn were mixed into the soil of a flower bed but not mixed in the soil of a second flower bed located on the other side of the house. After two weeks, the flowers in the first flower bed were much greener than those in the second bed. The most reasonable hypothesis to explain this would be that

   a. the first flower bed received more water.

   b. the grass molecules were broken down by decomposers and absorbed by the roots of the flowers in the first bed.

   c. the flowers in the first bed ate the grass clippings.

   d. the flowers in the first bed absorbed blades of grass through their roots.

   e. the flowers in the second bed absorbed a yellow substance through their roots.

9. The actual cause of the color difference of the flowers in the two beds mentioned in number 8 is not possible to determine without further testing because

   a. the flowers are not the same color.

   b. the weight of the soil was not measured.

   c. distilled water was not used to water the flowers.

   d. the grass was not mixed in the soil of both beds.

   e. the two flower beds differ in more than one way.

10. A small group of people are stranded on a barren island with a thousand bushels of wheat and one cow. To survive for the greatest length of time, the people should

    a. eat the cow, then eat the wheat.

    b. feed the wheat to the cow and drink the milk.

    c. feed the wheat to the cow, drink the milk, then eat the cow.

    d. drink the milk, eat the cow when milk production ceases, then eat the wheat.

11. Within any community, which would have the smallest total mass?

    a. producers

    b. primary consumers

    c. secondary consumers

    d. tertiary consumers

12. Which of the following does not recycle in an ecosystem?

    a. carbon

    b. oxygen

    c. water

    d. phosphorous

    e. energy

The following two items refer to the forest of the Pacific Northwest. Douglas firs, cedars, and hemlocks are the principle trees of the region. Characteristics of each are listed in the Table 12.2.

TABLE 12.2 OBSERVED CHARACTERISTICS

DOUGLAS FIR	CEDAR AND HEMLOCK
Seedlings die in shade.	Seedlings grow in shade.
Seedlings grow well on ashes.	Seedlings do not grow well on ashes.
Seeds are winged.	Seeds are not winged.

13. When an old Douglas fir tree dies in a dense forest, its place will be taken by

    a. Douglas fir seedlings only.

    b. cedar seedlings only.

    c. a mixture of cedars and hemlocks.

    d. hemlock seedlings only.

14. The effect of a fire in this area should be to increase the number of

    a. Douglas firs.

    b. cedars.

    c. hemlocks.

    d. pines.

The next two items (15–16) are based on Figure 12.13.

15. Which of the following processes is represented by the figure?

    a. matter flow

    b. succession

    c. natural selection

    d. spontaneous generation

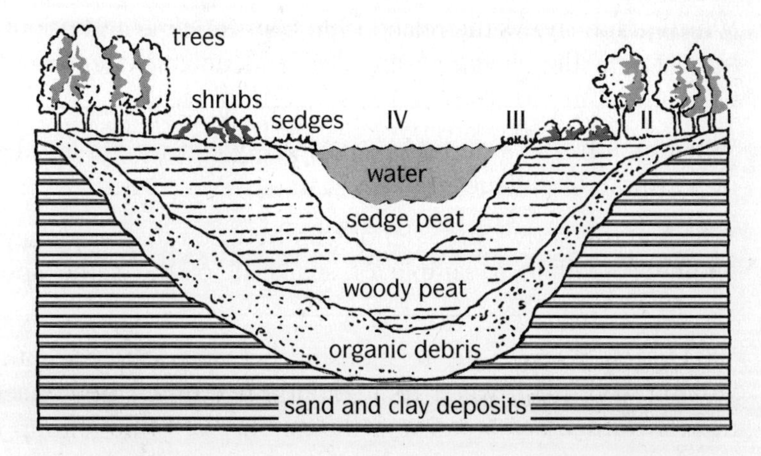

Figure 12.13 A cross-section of layers of the earth.

16. The next layer deposited at Position III probably will be
    a. clay.
    b. water.
    c. woody peat.
    d. sand.

Items 17–22 refer to the graphs in Figure 12.14.

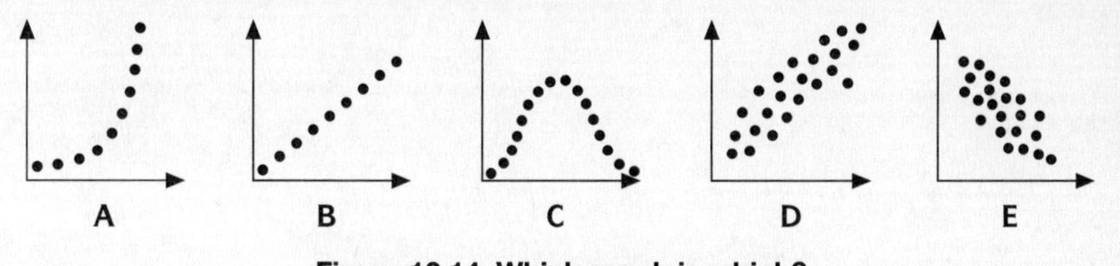

Figure 12.14 Which graph is which?

17. Which graph shows the relationship between length and weight of round brass rods that are all of the same thickness?

    ____

18. Which graph shows the relationship between age and the time it takes a sample of people (ages 25–65 years) to run one mile?

    ____

19. Which graph shows the relationship between time and population size of fruit flies living in an ideal and unlimited environment? ________________

20. Which graph shows the relationship between weight and height of a sample of college students? ________________

21. Which graph shows the relationship between the frequency and width of shell in a sample of snails all of the same species? ________________

22. Which graph shows the relationship between salt concentration varying from fresh water to a solution five times the concentration of normal sea water and the amount of brine shrimp eggs that hatch in those solutions? ________________

CHAPTER 13

THEORIES OF ORGANIC CHANGE

GETTING FOCUSED

- *Identify a theory of the origin of life.*
- *Understand organic evolution theory.*
- *Learn about Darwin's theory of natural selection.*
- *Learn about a synthetic theory of evolution.*
- *Evaluate how hypotheses about the origin and nature of life in the past can be tested.*

A THEORY OF THE ORIGIN OF LIFE ON EARTH

According to astronomers, the Earth formed from the condensation of a cloud of cosmic dust and gas between 4.5 to 5 billion years ago. Initially the Earth was a hot molten mass with no atmosphere. Gradually, small inorganic molecules such as H_2O, CO_2, CO, N_2, N_2S, and H_2 escaped from volcanic activity and collected above the Earth's surface to form a primitive atmosphere. We know about the escape of these molecules from present-day volcanic activity. Hydrogen cyanide gas (HCN) easily forms from these molecules and was also most likely present in the primitive atmosphere.

Presumably, these small molecules were mixed in the atmosphere and were then sparked by a variety of energy sources present on the primitive Earth. The energy sources consisted of solar radiation (visible light, ultraviolet light, X rays), lightning, and heat from the Earth's interior. When sparked, the molecules underwent chemical reactions, eventually to produce larger and more complex molecules, such as amino acids, fatty acids, purine bases, and simple sugars. These are the basic building blocks of the major types of molecules found in living things

Because the early atmosphere and the early oceans contained no free oxygen and no decomposer organisms, the basic building block molecules that formed were not broken down. Instead, they accumulated in the oceans over hundreds of millions of years. Those building block molecules were further concentrated in puddles on the beaches of ponds and lagoons, where the heat of the sun evaporated most of the water. Here the sun's energy caused the concentrated molecules to combine into more complex chains of molecules that could then be washed back into the oceans or lakes.

At this point, small cell-like droplets began to form from these chains of complex molecules. The droplets had a marked tendency to absorb and collect other molecules from the surrounding liquid. Some of the droplets contained particularly favorable combinations of molecules and thus increased in size until they became too large and broke apart, which produced new, smaller droplets similar to the originals. Those molecule-absorbing, "growing," and "reproducing" droplets gradually increased in complexity to produce the first true living cells on Earth some 3.5 billion years ago.

POSTULATES OF THEORY OF THE ORIGIN OF LIFE ON EARTH

1. Earth's primitive atmosphere consisted of small inorganic molecules, such as H_2O, CO_2, CO, N_2, N_2S, H_2, and HCN.
2. Inputs of energy from solar radiation, lightning, and the Earth's heat caused these molecules to combine into larger, more complex organic molecules, such as amino acids, purine bases, and simple sugars, which accumulated in the early oceans.
3. Sunlight shining on small puddles caused the further concentration and combination of organic molecules into chains of protein, carbohydrate, fat, and nucleic acid molecules.
4. Small droplets of these complex molecules formed that absorbed still more molecules to increase in size and eventually break apart to form more, similar droplets.
5. The small droplets increased in complexity to produce the first true living cells on Earth.

Testing a Postulate of the Theory

Postulate 2 of the theory argues that the input of energy from sunshine and such other naturally occurring sources as lightning was sufficient to cause simple inorganic molecules (i.e., those found to occur naturally in the nonliving environment) to combine spontaneously to form organic molecules (i.e., more complex molecules found in all organisms.) A number of experiments have been performed to test this postulate. The first and most well known of these experiments was conducted in 1953 by Stanley Miller, then a graduate student working at the University of Chicago (see Figure 13.1 on page 308).

What Miller and others did to test this postulate was to collect the gases thought to be present in the early atmosphere and spark them with electrical sparks. The electrical sparks simulated lightning strikes in the primitive atmosphere. After several days of sparking, the gases were condensed, collected, and analyzed to see what new molecules, if any, had been produced.

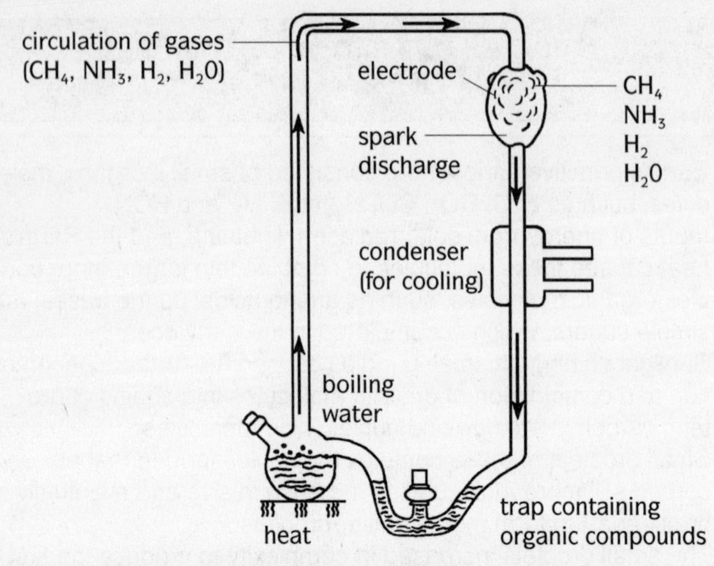

Figure 13.1 A diagram of Stanley Miller's experiment. The apparatus contained inorganic molecules thought to be present in the primitive atmosphere. Water was boiled to circulate the inorganic molecules past electrical sparks. The complex organic molecules that were formed were collected at the bottom. After W. T. Keaton, *Biological Sciences,* 4th ed. (New York: Norton, 1986).

The reasoning behind Miller's experiment and those that followed went as follows:

Hypothesis: *If*...basic organic molecules formed as a result of lightning striking simple inorganic gas molecules present in Earth's primitive atmosphere...

and...a collection of those simple inorganic gas molecules is sparked repeatedly by an electrical discharge...

Prediction: *then*...basic organic molecules should form.

Because these experiments invariably found that the basic organic molecules were produced in this way, as predicted, this postulate of the theory has been supported.

ORGANIC EVOLUTION THEORY

The theory of organic evolution (i.e., the idea that species change across time) represents a cornerstone of modern biology, as it conceptually ties together numerous and varied observations. The idea that species evolved had been suggested by others prior to publication of Charles Darwin's *The Origin of Species* (1858), yet Darwin was the first to clearly spell out major postulates of evolutionary theory.

POSTULATES OF EVOLUTIONARY THEORY

1. All life evolved from one simple kind of organism.
2. Each species, fossil or living, arose from another species that preceded it in time.
3. Evolutionary changes were gradual and of long duration.
4. Each species originated in a single geographic location.
5. Over long periods of time new genera, new families, new orders, new classes, and new phyla arose by a continuation of the kind of evolution that produced new species (equals adaptive radiation).
6. The greater the similarity between two groups of organisms, the closer is their relationship and the closer in geologic time is their common ancestral group.
7. Extinction of old forms (species, etc.) is a consequence of the production of new forms or of environmental change.
8. Once a species or other group has become extinct, it never reappears.
9. Evolution continues today in generally the same manner as during preceding geologic eras.

Testing Some Basic Postulates of the Theory of Evolution

Many ideas have been advanced to explain the origins of the vast diversity of life on Earth. Perhaps the three most prominent theories are the theory of evolution, the theory of special creation, and the theory of spontaneous generation. The experiments of Redi, Spallanzani, and Pasteur (see Chapter 6) did much to convince people that the theory of spontaneous generation (i.e., that new life arises spontaneously from nonliving matter, and that different types of nonliving matter give rise to different types of living things) is not a satisfactory explanation. Today this theory is no longer taken seriously by informed people.

The elimination from consideration of the theory of spontaneous generation leaves the theory of evolution and the theory of special creation. Did early organisms arise in the manner suggested by the theory of the origin of life presented in the previous chapter and then gradually evolve as suggested by Darwin and others? Or were they instead created by a divine and all powerful entity—a god? Of course, no theory is testable in its entirety. Instead, as we have seen, theories are tested postulate by postulate. We know what the postulates of the theory of evolution are, but what are the postulates of the theory of special creation? This question is not easily answered because there are many versions of the theory of special creation. To make the issue manageable, we will consider only the version that appears in the book of Genesis.

POSTULATES OF THE THEORY OF SPECIAL CREATION ACCORDING TO GENESIS

1. In the beginning, God created the heavens and earth.
2. God then created light and darkness to produce the first day.
3. On the second day, God created the sky to separate the water into two parts—water below the sky and water above the sky.
4. On the third day, God (a) gathered the water below the sky into one place to produce land and seas and (b) created seed and fruit-bearing plants to inhabit the land according to their kinds.
5. On the fourth day, God created two great lights (presumably the sun and the moon) and the stars.
6. On the fifth day, God created all the creatures of the sea according to their kinds and every bird according to its kind.
7. On the sixth day, God created all the creatures of the land according to their kinds, including a man in God's own image and a woman.
8. On the seventh day, God had finished his work and so rested.

Given the postulates of the theory of evolution (according to Darwin) and the theory of special creation (according to the book of Genesis), we are in a position to begin to try to decide which theory better explains how present day species arose. To test these theories, we must do what always must be done; that is, deduce some predictions from one or more of the postulates of the theories.

Before we discuss just what those predictions might be, we should first be very clear on what the two theories state. First, consider the theory of evolution. This theory states that all organisms arose from one simple kind of organism—most likely a single-celled organism that lived in shallow bodies of water (see the theory of the origin of life). Second, it states that more complex and varied life forms arose gradually over long periods of time. Third, it states that some kinds of organisms have become extinct. The Genesis theory of special creation paints a very different picture. According to this theory, complex life arose very quickly (all during a five day period). Also, since there is no mention of change, one can assume that organisms were created "according to their own kinds" in generally the same form as they appear today. Although there is no mention of extinction, perhaps the flood can be presumed to cause some extinctions. Thus the picture presented by the Genesis theory of special creation is one of rapid creation, no major changes in the basic forms of life, and possibly with some extinction, while the theory of evolution paints a picture of gradual development and major changes in life forms over vast periods of time, also with some extinctions.

Unfortunately, no one living today was an eye witness to these presumed events of the past, so we must rely on indirect evidence to test these theories. This evidence is contained in a record preserved in rock—the **fossil record.** A detailed discussion of the origin of the fossil record is too complex to go into here. It is enough to say that deposits of sand, silt, clay, and the like have, over vast periods of time, produced layers of sedimentary rock such as those found in the walls of the Grand Canyon. The oldest layers of rock are on the bottom and progressively younger layers are on the top. Embedded in these rock layers are fossils of organisms that lived in the past at different times. Thus, the fossil record presents a chronological history of life on Earth.

What do the two theories predict the fossil record should look like? Consider first the following lines of reasoning about the theory of evolution.

Hypothesis: *If*...organisms evolved over vast periods of time from simple to complex forms...

and...the record of organisms living in the past is examined in the fossil record embedded in layers of sedimentary rock...

Prediction 1: *then*...the lower (older) layers of rock should contain fossils of only a few very simple kinds of organisms;

Prediction 2: *then*...the higher (younger) layers of rock should contain fossils of progressively more complex and more diverse organisms.

On the other hand:

Hypothesis: *If*...organisms were created during a short period of time (five days) generally in the forms in which they appear today...

and...the record of organisms living in the past is examined in the fossil record embedded in layers of sedimentary rock...

Prediction 1: *then*...the lower (older) layers of rock should contain fossils of virtually all the complex and diverse kinds of organisms present today;

Prediction 2: *then*...the higher (younger) rock layers should contain fossils of the same complex and diverse kinds with the possible exception of those lost during the flood.

Thus the two theories lead to vastly different predictions about what the fossil record should look like. If the fossil record reveals a slow, gradual progression of changing life forms from few and simple to many and complex, then the theory of evolution has been supported. On the other hand, if the fossil record reveals an abrupt abundance of many complex forms that remain from lower to higher layers, then the Genesis theory of special creation has been supported. What then does the fossil record look like? The answer is that it looks just like it should according to the theory of evolution. No fossils are found in the oldest known rocks. Rocks that were formed 3–4 billion years ago contain fossils of only tiny, single-celled creatures. And progressively younger rocks contain progressively larger, more complex and variable life forms. The fossil record clearly reveals that the development of life on Earth has been a very, very slow and gradual process that has been going on for an extremely long time. The fossil record provides no evidence to support the theory of special creation as it is presented in Genesis.

DARWIN'S THEORY OF NATURAL SELECTION AND A SYNTHETIC THEORY OF EVOLUTION

In 1838, the Englishman Charles Darwin reached the conclusion that organisms change through time and for the next several years struggled to develop a theory to explain how this change could happen. Darwin's theory, called the theory of **natural selection,** was published in 1858 in his book, *The Origin of Species*. This book is without question the most important book in the history of biology because it presented a theory that organized and explained a greater number of observations about living things than any other book past or present. Darwin's theory of natural selection consists of many key ideas.

First, all species produce many more offspring in each generation than actually survive to reproduce. A single breeding pair of elephants, the slowest breeders of all animals, could produce a group of 19 million elephants in just 750 years if none of the offspring died before they were able to breed. The world would be overrun with elephants. Oysters are much more prolific. They have been known to produce over 100 million eggs in a single spawning. If all of the eggs developed into oysters. the oceans would be overrun with oysters in a few short years. This capacity for increase in numbers is referred to as biotic potential (see Chapter 12). Second, the biotic potential of organisms is seldom reached because of such factors as lack of food, space, or water; the environment may be too cold or too hot, or predation may occur. In short, factors within the environment limit the growth of populations. They are called limiting factors.

Third, the individual members of any given species differ from one another in many ways. With the exception of identical twins and single-celled organisms that reproduce by splitting, no two individuals are exactly alike. Humans differ in height, weight, eye color, hair color, body build, and so on. Other organisms differ in some obvious and some not so obvious ways. In other words, there is variation among the individuals in all populations.

Fourth, individuals with certain characteristics have a better chance of surviving and reproducing than individuals with other characteristics. When individuals compete for limited resources, or are subjected to predation, the stronger, faster ones—or sometimes the ones that are better able to hide—survive while the others die. There is a survival of the fittest. The ones that are not fit die and cannot reproduce.

Fifth, some of the characteristics that result in differential survival and reproduction are passed from parents to offspring (i.e., are heritable). Although characteristics that may be acquired during one's lifetime, such as a good suntan, large muscles from lifting weights, or a good golf swing, are not passed on to offspring through genes, many characteristics are. Good vision, height, and overall quickness are characteristics that are influenced considerably by one's genes and therefore may be passed from parent to offspring (see Chapter 7).

Sixth, the environments of most organisms have been changing through vast time periods. The geological record indicates that tremendous changes have occurred on Earth. Some areas that are now deserts were once rain forests. Some mountainous areas once lay at the bottom of inland seas. Some cold, snowy climates were once hot and tropical. Continents have moved and entire mountain ranges have been built and torn down. Changes such as these have been going on over billions of years in the past and continue to this day.

The consequence of these six factors is that favorable characteristics are passed from generation to generation through the genes and tend to accumulate in the population over time, while unfavorable characteristics decrease. This allows species to become better suited to changing environments and new species to arise. The entire process is referred to as natural selection because it is "nature" that selects which characteristics are favored and which are unfavored. This process differs from **artificial selection** (i.e., selective breeding of plants and animals with certain favorable characteristics by humans), primarily because nature does the selecting rather than people.

POSTULATES OF DARWIN'S THEORY OF NATURAL SELECTION

1. Populations of organisms have the potential to increase at a geometric rate.
2. In the short run, the number of individuals in a population remains fairly constant because the conditions of life are limited.
3. Individuals in a population are not all the same; they have **variations** (variable characteristics).
4. There is a struggle for survival so that individuals having favorable characteristics will survive and produce more offspring than those with unfavorable characteristics.
5. Some of the characteristics responsible for differential survival and reproduction are passed from parent to offspring (i.e., they are heritable). Hence there is a natural selection for certain favorable characteristics.
6. The environments of many organisms have been changing throughout geologic time.
7. Natural selection causes the accumulation of favorable characteristics and the loss of unfavorable characteristics to the extent that new species may arise.

These seven postulates taken together explain how variations in species can lead to changes in organisms when they are subjected to natural selection. At the time Darwin proposed his theory, he was not aware that new variations can arise by spontaneous changes in the genetic material (DNA). We now refer to these spontaneous changes as *mutations*.

Testing Postulates of Darwin's Theory of Natural Selection

When Darwin proposed his theory of evolution through natural selection he expected that it could be documented only thorough indirect evidence. The process of natural selection simply took too long for anyone to document directly, at least so Darwin thought. As it turned out, Darwin was wrong on this point.

During the nineteenth century, a light-colored variety of moths known as peppered moths (*Biston betularia*) was commonly found throughout much of the English countryside. The peppered moths flew at night in search of food and rested during the day on the trunks of lichen-covered trees and rocks. They were nearly invisible when seen against this similarly light-colored background. Until 1845 all of the peppered moths were light colored, but in that year one black individual was found near the newly industrialized city of Manchester. At the time, no one knew just what caused the moth, which otherwise looked very much like a peppered moth, to be black. Today we know that the black peppered moth, called *Biston carbonaria,* is the result of a mutation.

In the early and middle part of the nineteenth century, many of the English cities were being transformed from rural farming communities to industrial centers. With the industrial revolution came factories and pollution. Near factories, the once green countryside was becoming covered with a layer of black soot. The black soot covered tree trunks and rocks that were covered with light green lichens. This killed the lichens, leaving the trees black. During this time, more and more black-colored moths were being found. It was not long before 99% of the moths that were collected near industrialized cities such as Manchester were black. What had happened? Did the black soot cover the moths and changed them from light green to black? Although this seemed a good guess, the English biologist H.B.D. Kettlewell thought differently. He hypothesized and, through some very convincing experiments, demonstrated that the change was due to natural selection.

Although no one had ever seen a bird capture and eat a peppered moth, Kettlewell hypothesized just that. He believed that at first the light-colored moths were nearly invisible against the light-colored backgrounds. But when the countryside turned black due to the soot, the light-colored moths became visible to the birds and began being captured and eaten as before. Furthermore, because color was an inherited trait, the black color was passed to the offspring of the black moths. In this way, the number of black moths increased while the number of light-colored moths decreased. This continued until before long almost the entire population consisted of black moths.

To test his hypothesis that birds were actually selecting different colored moths depending on the color of the background, Kettlewell took both light-colored and black moths to two locations. One location was near Birmingham, an industrial area where nearly 90% of the moth population was black. The other location was near Dorset in an unpolluted countryside. Kettlewell then placed moths on tree trunks, and focused hidden cameras on them. By doing this, he was actually able to record birds capturing and eating the moths.

Kettlewell's reasoning was as follows:

Hypothesis: *If*...birds were actually selecting different colored moths, depending on the color of the background...

and...equal number of light- and dark-colored moths are placed on both light- and dark-colored tree trunks...

Prediction: *then*...more dark-colored moths should be captured on light-colored tree trunks and more light-colored moths should be captured on dark-colored tree trunks.

Kettlewell's film revealed that the capture was so swift that it was no wonder no one had ever observed it previously. On the black-colored tree trunks of Birmingham, when equal numbers of black- and light-colored moths were available, the birds captured 43 light-colored moths but only 15 of the black ones. Near Dorset, however, the birds captured 164 black moths but only 26 light-colored ones. Thus Kettlewell's experiments provided evidence that natural selection operates in nature just as Darwin thought it did.

During the 1950s and 1960s, evolution theory, the theory of natural selection, gene theory, and the process of genetic mutation were firmly enough established that biologists were able to combine these theories into one "synthetic" theory of evolution.

POSTULATES OF SYNTHETIC THEORY OF EVOLUTION

1. Each species is an isolated set of genes (a **gene pool**).
2. Evolution is the change of gene frequencies in the gene pool of a species or subspecies.
3. Individuals contain only a portion of the genes in its species gene pool, and the portion is different for each individual.
4. The particular combination of genes in a sexually reproducing individual is the result of the combination of genes from its parents, recombinations of its own genes, and mutation.
5. Separation that restricts the flow of genes between a subpopulation and its parent population is essential for the subpopulation to evolve into a new species.
6. Changes in gene frequency are the result of natural selection, migration, and random genetic changes such as mutation.
7. Evolution of a species may result in the change of a particular species across time (phyletic evolution), and/or an increase in the number of species (**adaptive radiation**).
8. Speciation (the process of forming biological species) is complete when variations between the evolving groups are large enough so that genetic exchange between the two groups cannot occur even if individuals of the two groups meet.
9. Mutation is the ultimate source of new genes in a gene pool.

QUESTIONS

1. Experiments such as the one described by Miller indicate that no special supernatural processes or powers are necessary to get simple inorganic molecules to combine into larger, more complex organic molecules. Apparently, all that is needed is for a number of these small inorganic molecules to accumulate in one place and that energy be supplied. This starts molecular collisions until the small molecules "spontaneously" combine. In a sense then, this theory of the origin of life involves the idea of spontaneous generation. But the experiments of Redi, Spallanzani, and Pasteur presented in Chapter 3 argued that the theory of spontaneous generation is wrong. Instead the theory of biogenesis is correct. Are the theory of biogenesis and the present

theory of the origin of life on Earth contradictory to each other or not? Explain.

2. Scientists do not claim that theories of special creation (i.e., explanations for the origin of life on Earth by acts of a supernatural, all-powerful "creator") are necessarily wrong. Rather, scientists consider such explanations to be outside of the realm of science. Why must all such theories lie outside the realm of science?

3. Does the existence of intermediate forms of life, as discussed in Chapter 7, support the theory of evolution or the Genesis theory of special creation? Explain.

4. Fungi are organisms that obtain nutrients by excreting digestive enzymes into the surrounding environment. They then absorb the dissolved nutrients directly through their cell walls. Six major classes of fungi have been identified. One class, called the fungi imperfecti, which contains the athlete's foot fungus and penicillium, contains over 35,000 species. Does the fact that the fungi imperfecti, a rather obscure group of organisms, contains over

35,000 different species support the theory of evolution or the Genesis theory of special creation? Explain.

____

____

____

____

5. Can you think of any piece of evidence that supports the theory of evolution that could not also be explained by the work of an all powerful creator? If not, what does this tell you about our ability to test theories that postulate the existence of an all-powerful creator? (See question 2 in Chapter 7.)

____

____

____

____

6. Kettlewell gathered other data designed to test Darwin's notion of natural selection. For example, he set out traps to capture moths in both the unspoiled lichen covered woodlands far from the cities and in the woodlands of blackened, soot-covered trees near the cities. The number of light and dark moths that he captured from each type of woodland are shown in Table 13.1. The numbers in the table indicate how many of each type of moth were captured in each type of woodland. Do the numbers reveal

TABLE 13.1 MOTH DATA

	LOCATION	
TYPE OF MOTH	LIGHT UNSPOILED WOODLAND	DARK SOOT-COVERED WOODLANDS
Light	324	144
Dark	3	1403

a correlation between the type of moth and the place they were captured? Do the results support or contradict Darwin's theory of natural selection? Explain.

___

___

___

___

7. Prior to the time Darwin proposed his theory of natural selection, a French scientist by the name of Jean Baptiste Lamarck proposed his own theory of how evolution occurs. His theory, known as the theory of the inheritance of acquired characteristics, consisted of these two basic postulates: (1) As any part of the body is used, it develops and enlarges; likewise, parts that are not used weaken, become smaller, or even disappear. (2) Any characteristics so acquired during one's lifetime can be passed on to the offspring. What evidence can you think of that supports postulate 1 and/or postulate 2? What sort of an experiment could be used to test postulate 2? What is the predicted result of the experiment based upon Darwin's theory of natural selection?

___

___

___

___

8. At the time Charles Darwin proposed his theory of natural selection, he knew nothing about Mendel's theory of genetics nor did he know about mutations. Why are genetic mutations important in the process of evolution? Similarly, suppose genes did not undergo mutation. What would be the consequences for the process of natural selection?

___

___

___

___

TERMS TO KNOW

adaptation
adaptive radiation
artificial selection
fossil record
gene pool
natural selection
variation

SAMPLE EXAM

1. Reproduction in the broadest sense is least necessary for
   a. survival of the individual.
   b. survival of the species.
   c. the continuation of life.
   d. the formation of new species.
2. From the point of view of evolution, what is the greatest advantage of sexual reproduction over asexual reproduction? Sexual reproduction produces
   a. a greater variety of organisms.
   b. a consistency of traits generation after generation.
   c. continuance of the species.
   d. fewer eggs to be fertilized.
3. Given that sexual reproduction is an evolutionary advantage for plants because it increases variability, how can the presence in some flowers of both male (pollen) and female (ova) gametes best be explained?
   a. Such plants have not evolved.
   b. The pollen and ova do not mature during the same time of year.
   c. Self-pollination results in more genetic diversity than cross-pollination in such plants.
   d. Genetic diversity increases due to environmental influences.

e. Evolution favors such flowers in rapidly changing environments.

The next four items refer to the following chart. The width of each screened area indicates the number of kinds of vertebrates.

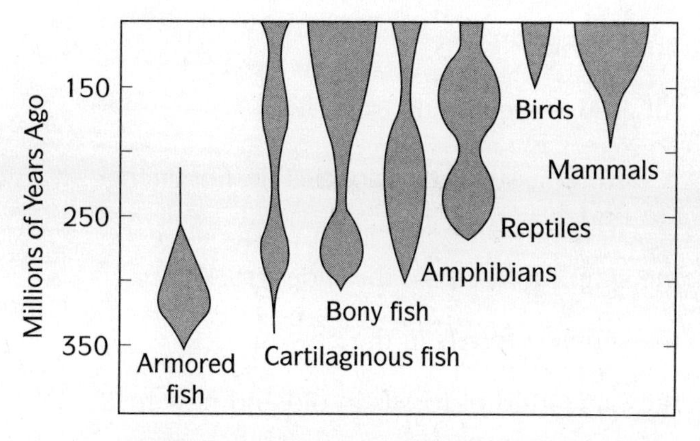

Figure 13.2 Fossil record of the vertebrates.

4. "After an animal group has originated, it tends to develop many kinds." This statement would be classed as

   a. a question.

   b. data.

   c. a prediction.

   d. an interpretation.

5. What appears to be the most recent group of animals to have come into existence?

   a. reptiles

   b. mammals

   c. bony fishes

   d. birds

6. The number of kinds of organisms in a group is a measure of successful adaptation. Which group of fish appears to have been the most successful?

   a. armored fishes

   b. cartilaginous fishes

   c. bony fishes

   d. All groups are equally successful.

7. If we were to assume that species do not change, we would expect to find

   a. the simplest fossils in the oldest rocks.

   b. the simplest fossils in the newest rocks.

   c. the same kind of fossils in old and new rocks.

   d. no fossils in any rocks.

8. If you constructed a time line in which one inch represented 500 years, about how long would your line be (in miles) if the present were at one end and the origin of eukaryotic life (2 billion years ago) were at the other end? Note: 1760 yards = 1 mile.

   a. 1 mile

   b. 10 miles

   c. 20 miles

   d. 60 miles

   e. 100 miles

Questions 9–15 refer to the following information about insects.

Assume that dark color is dominant to light color in a certain strain of insects. Ten pair of dark insects with heterozygous genotypes were released in a plot covering 5000 square meters of a white, sandy area. They could not escape from the area.

9. After two or three generations, what would be a predicted ratio of dark to light insects?

   a. 1 dark: 3 light

   b. 1 dark: 1 light

   c. 3 dark: 1 light

   d. They would all be dark.

   e. They would all be light.

Question: During the next few years, the actual proportion of dark insects decreased and the proportion of light ones increased. After a few years, the ratio of the insects was 3 light: 1 dark. Why did a difference develop in the numbers of dark and light insects?

Use the key to identify each of the next six statements relating to this question.

   a. a reasonable hypothesis in light of the data

   b. a hypothesis contradicted by the data

   c. not a hypothesis but a restatement of the data

   d. a prediction

10. There are more white insects now because light color is dominant to dark color. ____________________

11. After a few years, there were more light insects than dark insects. ____________________

12. The color of the light ones helped hide them from predators. ____________________

13. You could expect a continued shift in the gene frequency of the population. ____________________

14. Natural selection in this environment favored the light-colored insects. ____________________

15. The white sand caused dark insects to become light, and these then produced light insects. ____________________

GLOSSARY

abduction the mental process of generating hypotheses in which an explanation that is successful in one context is borrowed and applied as a tentative explanation in a new context.

abiotic factor nonliving environmental factor or agent that directly affects living things; such as water, wind, temperature, gases, and so forth; opposite of *biotic*.

acid any type of molecule that releases hydrogen ions (protons) when dissolved in water, producing a pH of less than 7 and having a sour taste; opposite of a base.

activation energy the amount of energy needed to start a chemical reaction.

active transport a cellular process that requires energy, which moves molecules across a cell membrane (from a low concentration to a high concentration) of those particular molecules.

adaptation in evolution, any genetically controlled characteristic that increases an organism's ability to survive and reproduce in its environment.

adaptive radiation the evolution of several new species from one ancestor. The new species have evolved to fill specific niches; also referred to as divergent evolution, speciation, or macroevolution.

adenosine diphosphate an adenosine phosphate molecule containing two phosphate groups.

adenosine monophosphate an adenosine phosphate molecule that contains one phosphate group.

adenosine phosphate a nucleotide that contains adenine, ribose (a sugar), and either one, two, or three phosphate groups involved in the storage and transfer of energy in cells.

adenosine triphosphate an adenosine phosphate molecule that contains three phosphate groups.

air pressure the collective push (force) of the moving molecules in air on a surface. The amount of air pressure depends on the speed of the molecules and their mass. More molecules of greater mass produce greater pressure.

allele alternate form of a single gene coding for the same characteristic; located at the same position (locus) on homologous chromosomes.

amino acids molecules that contain an amino group ($-NH_2$) and a carboxyl group (-COOH), both bonded to the same carbon atom; the "building blocks" of proteins.

animal an organism that obtains needed energy (food) by ingesting (via a mouth of some sort) other organisms. One of the major kingdoms in taxonomic systems. Animals are the consumer organisms in biological communities.

anthropomorphism attributing human characteristics to non-human entities.

artery vessel that carries blood away from the heart and toward tissues and cells; opposite of vein.

artificial selection the selective breeding of animals and plants for the purpose of obtaining offspring with certain desired characteristics.

assumption a statement taken to be valid (i.e., correct), usually for the sake of constructing an argument.

atom the smallest unit into which a material can be divided and still retain the chemical characteristics of that material.

base any type of molecule that releases hydroxide ions (OH^-) when dissolved in water, reducing the hydrogen ion concentration and producing a pH greater than 7; opposite of acid.

biodegradable describing materials such as paper made of complex molecules that can be broken apart (decomposed) by decomposer organisms (bacteria and fungi); opposite of *nonbiodegradable.*

biogenesis the idea that new living things can come only from other living things.

biological community all the organisms that live and interact in a particular area or habitat.

biological magnification a process in which certain retained molecules become progressively concentrated at each higher trophic level in a food chain; for example, nonbiodegradable herbicides like DDT, or heavy metals.

biology the study of the nature of living things and the nature of their interactions with their environments. The term *biology* may apply to the ways in which people study living things (the process of doing biology) and/or to the knowledge gained by that process (the product of doing biology).

biomass the combined weight of all organisms in a particular area.

biome worldwide grouping of complex communities characterized by unique vegetation and climate; for example, tropical rainforests, grasslands, and deserts.

biosphere the part of Earth that contains living things, including land, air, and water; all the Earth's biomes.

biotic factor describing a living environmental factor (i.e., organisms) that directly affect other organisms; for example, a fungus living on the leaves of a tree is a biotic environmental factor; opposite of *abiotic*.

biotic potential the maximum growth rate a population can experience under conditions ideal for its growth.

blastula the developmental stage of an animal embryo between initial cell division and gastrulation; typically a hollow ball made of a single layer of similar cells.

Calvin cycle the reactions of photosynthesis that can take place without light (the light-independent reactions), in which carbon dioxide molecules are combined into larger molecules, such as glucose.

carbohydrates a class of molecules found in living things, containing carbon, hydrogen, and oxygen usually in a 1C:2H:1O ratio; examples are sugars, starches, cellulose, and glycogen.

carnivore an animal that eats primarily meat.

carrying capacity the maximum number of individuals of a specific population that can be supported by the available resources of that population's environment.

catalyst any molecule or compound that regulates the conditions at which a chemical reaction occurs but does not become part of the end product of the reaction. Enzymes are catalysts that regulate the reactions in cells.

causal question a question that asks why, what is the cause of a specific event or set of related events. What condition or conditions provoked a specific result?

cell differentiation see **differentiation.**

cell membrane a thin, double layer of phospholipid and protein molecules that surrounds the cytoplasm of all cells and controls the movement of materials into and out of the cell; sometimes referred to as *plasma membrane.*

cell replication see **meiosis** and **mitosis.**

cell wall a relatively rigid layer of cellulose and lignin molecules lying outside the cell membrane of all plant cells, some protists, and most prokaryotic cells.

cells the basic structural and functional unit of all living things; usually microscopic in size.

cellular respiration a complex sequence of chemical reactions that take place inside cells when oxygen is present and that derive energy (ATP) from the breakdown of relatively large molecules, such as carbohydrates, fats, and proteins.

centromere a constricted region, usually at the center of chromosomes, at which recently duplicated chromosomes are held together; spindle fibers are also attached here during mitosis and meiosis.

characteristic observable, structural, functional, or behavioral feature or attribute of an object or organism; synonymous with *property* or *trait.*

chemical bonds attractive force that hold atoms together. There are three major types of chemical bonds:

ionic bonds A relatively strong chemical bond formed as a result of the attraction between oppositely charged ions.

covalent bonds A relatively strong chemical bond formed by the sharing of electrons between atoms.

hydrogen bonds A relatively weak chemical attraction formed between a hydrogen atom that possesses a partial positive charge and an atom (usually oxygen or nitrogen) that possesses a partial negative charge.

chemical energy the energy contained in chemical bonds that can be "released" when those bonds are broken.

chemical reaction interaction in which the bonds that hold the atoms of a molecule or compound together are broken or new bonds are formed to create new molecules or compounds.

chlorophyll any one of several green pigment molecules necessary for photosynthesis.

chloroplast organelle inside plant cells that contain chlorophyll molecules that function in photosynthesis.

chromosome cigar-shaped bodies of condensed DNA molecules found in the nucleus of cells; usually visible only during cell division.

climax community the final, stable stage in ecological succession; dominated by an interacting group of species that tend not to be replaced by other species.

combinatorial thinking thinking pattern used to generate all possible combinations of real or imagined objects, events, or situations.

conclusions statement that summarizes the extent to which hypotheses and/or theories have been supported or contradicted by the evidence.

constant a characteristic (property) in a collection of objects, events, or situations that does not differ from one to the other; for example, all the jars in his experiment were filled with 100 ml of water, so the amount of water is a constant.

consumer an organism that obtains its energy by directly ingesting (eating) other organisms (i.e., animals). Animals that eat plants (herbivores) are primary consumers, while animals that eat other animals (carnivores) are secondary or tertiary (i.e., second or third order) consumers.

controlled experiment an experiment in which the values of only one independent variable are allowed to change. The values of other independent variables are held constant or "controlled." Controlled experiments are central to science because they allow you to separate the possible effects of one variable (one possible cause) from the possible effects of others.

correlational thinking the thinking pattern used to identify and/or solve quantitative problems that involve the possible covariation of the values of two variables.

crossing over the exchange of segments of chromosomes between recently duplicated homologous chromosomes during cell replication that results in gene mixing and provides an essential source of genetic variation in sexually reproducing organisms.

cytokinesis division of the cytoplasm of a parent cell into two daughter cells.

cytoplasm everything inside cells, inside the cell membrane, and outside of the nuclear membrane.

dark reaction see **light-independent reactions.**

decomposer an organism that obtains its food supply (energy) by absorbing the molecules of living or nonliving organisms through its cell membranes (bacteria and fungi).

deduction the mental process of joining a hypothesis and/or generalization with a proposed experiment and/or observation to allow the derivation (development) of an expected outcome(s). The process of deduction follows the If...and...then form.

density dependent factor environmental factor that affects a population as a function of the size of the population. For example, as a deer population increases, its food supply may not. Therefore, the amount of food available will become more and more of a factor in limiting the size of the deer population.

density independent factor environmental factor that affects populations in ways that are independent of the size of the population. For example, a hard freeze might kill all of the lantana bushes in an area regardless of how many lantanas are present.

deoxyribonucleic acid (DNA) the complex molecule that functions as the genetic material (the genes) of all organisms; composed of two complimentary strands of nucleotides coiled together into a double helix.

dependent variable the variable in a controlled experiment whose values may vary in response to changes in the values of some independent variable. Sometimes referred to as the outcome variable, the effect.

differentiation the irreversible process of similar and unspecialized cells becoming different and specialized during developmental stages of multicellular organisms.

diffusion movement of atoms, ions, or molecules from regions of high concentration of these atoms, ions, or molecules to regions of low concentration as of result of random collisions of those particles with them and with the particles through which they are moving.

dominant (allele) the gene of a pair of genes that is "expressed" in the phenotype by "masking" the action of the other gene of the pair (the recessive gene/allele) of the same pair. Each gene of a pair is sometimes referred to as an allele.

ecological niche a description of an organism's "occupation" within its community; how an organism interacts with its environment.

ecosystem a community of organisms along with all associated abiotic (physical) factors that affect that community.

ectoderm in an animal embryo, the outermost of the three embryonic cell layers that gives rise to the nervous system and the outer layer of skin.

egg a female cell produced by meiosis that contains only half of the normal number of chromosomes (the female gamete).

electron transport chain the third and final stage of cellular respiration in which molecules of ATP are manufactured using the energy generated by the movement of electrons through a series of molecules located on the inner membrane of mitochondria.

electron primary subatomic particle with a negative charge that travels in orbits around the nucleus of an atom.

emergent property unique property of nonliving objects, organisms, or systems that arises as a consequence of novel arrangements of their parts. Emergent properties of living things and systems arise during the evolutionary development of species and the embryological and intellectual development of organisms.

endoderm the innermost of the three embryonic cell layers of animals that develops into the stomach, intestines, urinary bladder, and respiratory tract.

endothermic reaction a chemical reaction that requires an input of energy to proceed; opposite of exothermic reaction.

energy motion or the capacity to make things move or do work.

environment the sum total of an organism's environmental factors, an organism's surroundings.

environmental factor any factor external to an organism that may affect it in any way (e.g., amount of water, temperature, other organisms).

enzyme protein molecule that acts as catalyst to control the speed and direction of chemical reactions within living organisms without being used up by the reactions.

eukaryotic a type of cell that contains membrane-bound organelles such as a nucleus and mitochondria. All plants, fungi, and animals have eukaryotic cells; opposite of *prokaryotic*.

evidence actual observations of nature that can be compared with expected observations (predictions) to allow the test of alternative hypotheses and/or theories.

evolution change in the characteristics of members of a population due to changes in their gene frequencies from generation to generation.

excretion the elimination of nitrogenous wastes, such as urine.

exothermic a chemical reaction that releases energy as it proceeds; opposite of endothermic reaction.

experiment a set of manipulations and/or specific observations of nature that allow the test of hypotheses and/or generalizations.

exponential growth a pattern of population growth characterized by a steeply climbing, almost vertical curve, representing unchecked population growth.

extinction the permanent loss of all individuals of a species.

fact a statement concerning a direct observation of nature that is so consistently replicated (repeated) that virtually no doubt exists as to its validity.

fermentation cellular respiration that takes place when oxygen is not available to the cell; the reaction breaks apart relatively large molecules (e.g., carbohydrates, fats, proteins) to produce ATP molecules and alcohol or lactic acid molecules. Fermentation is less efficient than respiration because the large molecules are not fully broken apart when oxygen is not available.

fertilization the joining of egg and sperm cells (the female and male gametes) that results in the zygote, the first cell of the offspring individual.

food chain a sequence of feeding relationships starting with producers, moving to successive levels of consumers, and finally ending with decomposers; for example, seeds are eaten by mice that are eaten by a snake that is eaten by a hawk that dies and is eaten by bacteria and fungi.

food web the total of all interconnected food chains in a biological community.

fossil the remains of living beings preserved in rock.

fungus (plural, fungi) a type of multicellular organism that produces the enzymes that promote the breakdown of the relatively large molecules found in the bodies of dead organisms. The molecules are then absorbed into the cells of the fungus to be used as food. Fungi are decomposer organisms.

gamete a sperm (male) or egg (female) cell that contains half of the normal set of chromosomes. When egg and sperm unite during fertilization, a new cell (the zygote) is formed with a full complement of chromosomes.

gastrula an embryonic stage of animal development following the blastula stage, during which the cell layers of the endoderm, mesoderm, and ectoderm are initially formed.

gene pool all the genes of all the individuals in a population.

gene the basic hereditary unit, composed of a specific sequence of nucleotides on a DNA molecule on a chromosome. One gene typically codes for one protein or RNA molecule.

genotype the complete genetic makeup of an organism, whether or not it is expressed in the organism's phenotype.

guttation the expulsion of water from specialized regions of leaves as a result of root pressure.

habitat the place where an organism lives; an organism's "address."

herbivore an animal that eats plants, a primary or first-order consumer.

homeostasis the maintenance of a relatively stable (constant) internal environment of the cell or body by means of self-regulatory mechanisms, such as blood pressure or pH.

homologous chromosome in the nucleus of cells, one of a pair of chromosomes that usually resemble each other in length, shape, and the genes they contain. Typically, one chromosome of a homologous pair comes from the male parent and the other from the female.

hypothesis a single proposition intended as a possible explanation for an observed phenomenon, that is, a possible cause for a specific effect. A newly generated hypothesis that has yet to be tested is sometimes referred to as a speculative hypothesis. One that is in the process of being tested is sometimes referred to as a working hypothesis. A hypothesis that has been tested and repeatedly supported may take on the status of a belief, that is, provisionally (temporarily) valid or correct.

hypothetico-deductive reasoning the reasoning pattern used to test hypotheses via the imagination of some experimental and/or correlational situation and the deduction of its logical consequence(s). Hypothetico-deductive reasoning follows the linguistic if…and…then…therefore form of argumentation.

independent assortment during meiosis, the genes on separate chromosome pairs travel (assort) independently of each other in the production of gametes. The result of independent assortment is that genes located on separate pairs of chromosomes are inherited independently of each other.

independent variable the variable in a controlled experiment whose values are varied to see if that variation causes a change in the outcome (the dependent variable) of the experiment. If it does, the independent variable has been found to be a cause; sometimes referred to as the input variable, the cause.

induction the mental process of deriving (developing) general statements from a limited set of specific observations.

instincts any of the specific behaviors of organisms that are primarily the product of their genetic makeup. Instincts are not learned.

interspecific competition competition among members of different species for a limited resource such as food, shelter, or territory.

intraspecific competition competition among members of the same species for a limited resource such as food, shelter, or territory.

ion an atom or group of atoms that possesses a net positive or net negative electrical charge.

kinetic energy energy associated with the motion of objects as opposed to "stored" or potential energy.

Krebs cycle the stage of cellular respiration that completes the breakdown of glucose into carbon dioxide; occurs in the mitochondria and in the cytoplasm of some bacteria.

law a general statement that summarizes a pattern of regularity detected in nature, that is, the manner or order in which a set of natural phenomena occur under certain conditions. (Note that a satisfactory explanation for the pattern may or may not exist.)

learning the process of acquiring new behavioral patterns due to experience.

life the characteristic of being alive as opposed to nonliving or dead. Although there is no absolute distinction between the living and the nonliving, living objects usually are characterized by the ability to move, grow, reproduce, respond to stimuli, and exchange materials with their environment; they are chemically complex, highly organized, and consist of one or more structural and functional units called cells.

light-dependent reactions the first stage of photosynthesis that occurs on the thylakoid membranes of chloroplasts and converts light energy into chemical energy by the manufacture of ATP and NADPH molecules.

light-independent reactions the second stage of photosynthesis, in which carbon dioxide molecules are assembled into larger molecules, like glucose, using energy from the first stage (light-dependent reactions) of photosynthesis; these reactions can occur whether or not light is present; also referred to as *dark reactions*.

limiting factor any environmental factor that prevents an organism from living in a specific habitat or restricts its population size. Such factors include lack of food, too high or too low a temperature, and too much or too little light.

lipid a type of complex molecule composed primarily of hydrogen, oxygen, and carbon atoms that are insoluble in water; includes oils, fats, phospholipids, waxes, and steroid hormones.

meiosis a type of nuclear division that occurs only in organisms that reproduce sexually; involves two successive nuclear divisions, resulting in the number of chromosomes being reduced to half the normal number; one single cell produces four cells (gametes or spores) with half the initial number of chromosomes. Meiosis may also be referred to as *reduction division*.

mesoderm one of the three embryonic cell layers, which develops into all connective tissues, muscle, blood, kidneys, and several other structures; lies between the ectoderm and endoderm.

metabolic rate the overall rate at which the metabolism of an organism occurs.

metabolism the sum total of all chemical reactions occurring in an organism.

mitochondrion (plural: mitochondria) an elongated or spherical membrane-bound organelle in which cellular respiration takes place.

mitosis a type of nuclear division involving the replication and distribution of a complete set of chromosomes identical to both daughter cells.

mnemonic device words, phrases, or sentences that are associated with difficult to recall words.

molecule a particle that results from the joining of two or more atoms.

mutation a change in the kind, structure, sequence, or number of the component part of a DNA molecule; that is, a change in one or more of an organism's genes.

natural selection the process in which certain organisms that are better suited to live in a particular environment are able to survive and pass on their genes to the next generation. Natural selection is one of the most important processes that brings about evolutionary change.

neutron a primary subatomic particle with no electrical charge; found with one or more protons in the nucleus of an atom.

nucleic acid a large molecule made up of nucleotide subunits; the two types are *d*eoxyribo*n*ucleic *a*cid (DNA) and *r*ibo*n*ucleic *a*cid (RNA).

nucleotide a type of molecule that consists of a five-carbon sugar (either ribose or deoxyribose), a nitrogenous base, and a phosphate group; a "building block" of nucleic acids.

nucleus a membrane-bound organelle that contains chromosomes, often referred to as the control center of the cell; the central region of an atom that contains protons and neutrons.

omnivore an animal that eats both plants and animals, such as humans.

optimum range the range of values of a particular environmental variable (factor) that are most helpful to the survival and well-being of an organism.

organ system a collection of two or more organs that function together to carry out one or more processes necessary for the survival of the organism (e.g., digestive system, respiratory system).

organ a structure of definite form and function composed of more than one tissue (e.g., eye, kidney, lung, stomach, leaf, root).

organelle a specialized membrane-bound structure within cells that performs a specific function; for example, the nucleus, endoplasmic reticulum, and mitochondrion.

organic characteristic of, pertaining to, or derived from living things (organisms).

organism a living thing composed of one or more cells.

osmosis the diffusion of water through selectively permeable membrane; forms a region of higher concentration of water to a region of lower concentration.

pH (scale) (potential of hydrogen) a measure of the concentration of hydrogen ions (protons) in a solution. The pH scale runs from 0 to 14, with basic solutions having a pH greater than 7, and acidic solutions (many hydrogen ions) having a pH less than 7; pH 7 is neutral. See **acid** and **base.**

phenotype the sum total of an organism's characteristics that are dependent upon its genes or its "genotype."

phloem food-conducting tubes found in vascular plants.

phosphate group a group of seven atoms—one phosphorous, four oxygen, and two hydrogen—that bond with other groups of atoms to form molecules important in the storage and transfer of energy in cells. See **adenosine phosphate.**

photon an elementary "particle" or unit of light energy that cannot be further subdivided; also known as a *quantum.*

photosynthesis the cellular process in which light interacts with molecules of chlorophyll to initiate the manufacture of carbohydrate molecules from carbon dioxide and water molecules.

plants a major subdivision (a kingdom) of multicellular organisms capable of manufacturing their own food by using raw materials and light energy; the producers in biological communities.

polar covalent bond a relatively strong chemical bond in which the electrons are shared unequally between two atoms, resulting in the formation of a polar molecule. (See **polar molecule.**)

polar molecule a molecule with an unequal distribution of electrons that causes one part of the molecule to have a positive charge and the opposite part to have a negative charge, such as water.

population a group of organisms of the same species that inhabit a specific area and freely interbreed with one another.

postulate a statement that, when taken together with one or more other statements, attempts to explain a set of related phenomena (a theory).

potential energy energy "stored" in an object due to its position or the position of its parts relative to one another.

prediction a statement that represents a reasonable consequence (expected outcome) of an experiment and/or observation if the hypothesis and/or generalization under consideration is assumed to be correct. A prediction is the result of the process of deduction. The term *prediction* also refers to the outcome of the process of "guessing" about future events based upon an extrapolation (prediction) of a pattern of regularity detected in past events.

preformation the hypothesis that successive generations of individuals were present in tiny preformed versions in previous generations. The hypothesis has not been supported by the evidence.

probabilistic thinking the thinking pattern used to identify and/or solve problems that involve quantitative probabilistic relationships.

producer any organism in a biological community that uses raw materials and light energy to manufacture its own food. Directly or indirectly, producers supply all of the food for other organisms in the community. See **plants.**

prokaryote single-celled organism that has no nucleus or other membrane-bound organelles; the bacteria and blue-green algae.

proportional thinking the thinking pattern used to identify and/or solve problems that involve quantitative proportional relationships.

protein a class of large, complex molecules made of amino acid subunits; primary components of all cells.

protist single-celled eukaryotic organism. Protists are one of the major subdivisions of organisms. Members of the Protista kingdom.

protons a primary subatomic particle with a positive charge; found in the nucleus of an atom along with one or more neutrons; a hydrogen ion in solution.

radioactive referring to certain materials that give off energy in the form of particles and/or waves.

reasoning the acquisition of knowledge and/or new behaviors by thought.

recessive the gene of a pair of genes that is not "expressed" in the phenotype because its potential action is "masked" by the action of the other gene of the pair (the dominate gene/allele).

reflex a simple, repeatable movement of an organism or some part of the organism, provoked by some external stimulus.

respiration the process of gaseous exchange (usually oxygen and carbon dioxide) between an organism and the environment. (See **cellular respiration.**)

ribonucleic acid (RNA) a nucleic acid that functions mainly in the process of protein synthesis; composed of a single strand of nucleotides. RNA is also the genetic material of a large number of viruses. A form of RNA, called messenger RNA, carries genetic information from the nucleus to the ribosomes in the cytoplasm. Another form of RNA called transfer RNA carries amino acids to the ribosomes. The ribosomes are where amino acids are joined to form protein molecules.

root pressure pressure in root xylem cells developed by the roots as a result of osmosis; causes guttation of water in leaves and cut stems.

scavenger in biological communities, an animal that eats dead organisms (via a mouth). In this sense, scavengers act like decomposer organisms, as opposed to the "real" decomposers, the fungi and bacteria, which absorb dead organisms.

scientific method the collection of various activities that people use in their attempt to describe and explain natural phenomena accurately. The scientific method involves, but is not limited to, the following activities:

1. Raising both descriptive and causal questions about nature.
2. Inventing alternative explanations for what is observed.
3. Developing experimental and/or observational means to allow the explanations to be tested.
4. Deducing the expected outcome(s) of experimental and/or observational tests, assuming the explanation is correct.
5. Gathering and analyzing data to determine the extent to which the data are in agreement with the expected outcome(s).
6. Generating and communicating conclusions regarding the relative support or lack of support obtained for the alternative explanations.

selectively permeable the property of biological membranes that allows some materials (water and small ions) to pass through more easily than other materials (large molecules); also referred to as *differential permeability* or *semipermeable.*

solute any material that is dissolved in a solvent to form a solution.

solution any homogeneous liquid composed of two or more materials.

solvent any liquid that acts as a dissolving agent. Water is considered the "universal" biological solvent.

special creation the theory that explains the origin of living things by a supernatural act of a creator.

speciation the evolution of a new species; also referred to as macroevolution.

species a group of organisms that share enough characteristics to be able to mate and produce fertile offspring.

sperm/pollen male gamete in sexually reproducing organisms with a tail-like structure for propulsion. Sperm carry only half of the normal complement of chromosomes/genes.

spindle fibers (apparatus) threadlike structures visible in dividing cells that provide the framework on which chromosomes move during mitosis and meiosis.

spontaneous generation the theory of the origin of living things from natural and internally provoked activities of nonliving materials.

succession the gradual change in the species that inhabit and/or dominate a community at a particular site. Succession usually ends with a climax community in which the species no longer change, provided the climatic conditions remain the same.

taxis (plural: taxes) a continuous movement, either toward or away from a specific environmental stimulus, such as light.

teleological attributing a sense of purpose to living things other than humans.

temperature a measure of the amount of motion of the atoms and/or molecules of a particular material. The greater the motion, the greater the temperature.

theory a collection of general statements (postulates/assumptions) that when taken together attempt to explain a set of related phenomena.

thinking pattern a sequence of mental operations used to organize and/or analyze information.

thylakoids flattened pancakelike structures within chloroplasts in which photosynthetic pigments are arranged; the site of the light-dependent reactions of photosynthesis.

tissue a group of similar cells with a common structure and function.

transpiration the loss of water from the portions of plants exposed to the air. Most transpiration occurs through tiny holes on leaves called *stomata*.

trophic level a feeding level in a biological community, such as producers or first- or second-order consumers; a level in the flow of energy and biological materials through a community.

tropism a growth response (movement) toward (positive) or away (negative) from an external stimulus.

value the specific measure of a variable characteristic; for example, specific values for the variable of height in a sample of students might be to 5'11", 4'10", 6'2"; the specific values for the variable of color of light in a photosynthesis experiment might be blue light, red light, and green light.

variable a characteristic (property) in a collection of objects, events, or situations that differs from one to the other (e.g., in a group of college students, their heights differ). Compare the term variable to the opposite term *constant*.

variation anatomical, physiological, or behavioral differences within a population of organisms due to environmental conditions, or the genetic make up of the individual organisms.

veins a thin-walled vessel that carries unoxygenated blood toward the heart; opposite of artery.

virus a noncellular structure composed of a nucleic acid (DNA or RNA) core and a protein coat. Viruses can reproduce only if they infect a host cell.

vitalism the hypothesis that living things differ from nonliving things primarily due to the presence of a special living spirit or force.

xylem vascular tissue that conducts water and minerals from the roots to the rest of the plant; also functions in mechanical support.

zygote a resulting cell from the fusion of sperm and egg (or equivalent haploid cells); a fertilized egg; the first cell of the next-generation individual that contains a complete complement of chromosomes/genes.

APPENDIX: ANSWERS TO QUESTIONS AND SAMPLE EXAMS

CHAPTER 2

Pages 48–51

1–3. Answers will vary.

4. a. Turn over the E card, because an odd number on the other side would break the rule.

   Do not turn over the K card, because we have no rule about consonants.

   Do not turn over the 4 card, because neither a vowel nor a consonant will break the rule.

   Turn over the 7 card, because a vowel on the other side would break the rule.

   b. Can't tell from the clues because no information about this possible trip as given can be inferred

   c. Yes. They can travel from Fish to Bird with a stop at Bean.

   d. No. *If* travel between Fish and Snail were possible, *then* travel would be possible from Bird to Snail (via Bean and Fish), *but* this is not possible, according to the second clue. *Therefore,* travel between Fish and Snail is not possible.

CHAPTER 3

Page 56

1. They are brief, global, incomplete, incorrect, and non-analytical.
2. They are complete, aware of variables and the need to control, more analytical, use assumptions, uses hypothetico-deductive reasoning, aware of probabilities.

Page 58

1. The degree of completeness and systematic approach
2. Type B responses were more complete, more systematic.

Page 60

1. Type B responses used proportions, the type A responses add and subtract. The type A need to see the frogs to count them. The type B were able to go beyond observations and deal with estimation, the possible, and reason using assumptions in a hypothetico-deductive fashion.

Pages 72–77

1. Descriptive response is 8. $4 + 2 = 6$; therefore, $6 + 2 = 8$.

   Hypothetical response is 9. $4/6 = 6/x$, $4x = 36$, $x = 9$. (proportional reasoning)

2. Descriptive response is 9. $11 - 2 = 9$

   Hypothetical response is 7 1/3. $4/6 = x/11$, $6x = 44$, $x = 7\ 1/3$ (proportional reasoning)

3. Descriptive response, use string 1 because it has the most weight and is long.

   Hypothetical response, use strings 1 and 2, because they vary in length but have same weight amount; therefore, if you get a difference in time of swing, we know it must be due to string length (control of variables).

4. Descriptive response, use string 1 or 2, because they have most weight.

   Hypothetical response, use strings 1 and 3, because they vary in weight but have same length string; therefore, if you get a difference in time of swing, we know it must be due to amount of weight (control of variables).

5. Descriptive response: Answers will vary. Hypothetical response: b, gravity but not to red light. Tube III shows flies at top. Because light appears to be the same in both ends they must be responding to gravity. A comparison of tubes II and IV show no

real difference in fly distribution. If they responded to red light (go to or away from) then most in tube II should be either in light or dark end (control of variables, probabilistic reasoning).

6. Descriptive response: Answers will vary. Hypothetical response: c, both blue light and gravity. Tube III shows a response to gravity as in question 5. Tube II shows that flies move toward blue light and away from dark end of tube (control of variables, probabilistic reasoning).
7. Descriptive response: Answers will vary. Hypothetical response: 7/21 = 1/3 because 7 of the total number of pieces (21) are red or blue circles (4 red + 3 blue = 7) (probabilistic reasoning).
8. Descriptive response: Answers will vary. Hypothetical response: 15 combinations (combinatorial reasoning).
9. Descriptive response: Answers will vary. Hypothetical response: No, because the ratio of large to small fish is the same (4/12 = 3/9) in both groups of fish (i.e., those with narrow stripes and those with wide stripes) (correlational reasoning).

CHAPTER 4

Pages 87–88

4. 1/36 = 2.8%, 6/36 = 1/6 = 16.7%
5. a. 25; 4/5

   b. 80%

   c. 80%
6. a. 5/75 = 6.7%

   b. 6.7%
7. a. 23/28 = 82%

   b. 18%

   c. No, because this class is not likely to be representative of the school's population with respect to the sex of the students.

8. Answers will vary. For example: Assume that the probability of finding a needle in a 1-cubic-foot haystack in one hour is about 1. Assume that the probability decreases with an increase in size of stack (i.e., the probability would be 1/2 with a 2-cubic-foot stack). A five-foot-tall stack would contain about $(5 \times 5 \times 5) \div 2 = 62.5$ cubic feet of hay. Therefore, the probability would be about 2 out of 62.5. Variables may include visual acuity, size of needle, shape of haystack, and so forth.

Pages 101–104

1. a. Yes, it shows a correlation between the speed the auto was traveling and the distance the auto traveled on one gallon of gasoline. The slower the speed of the auto, the further it traveled, and likewise the faster the auto traveled the shorter the distance the auto would cover.

   b. Point number 5 may be an exception.

   c. The car could have been a different make, wind could have played a role, a different type of fuel could have been used, or some other force might have acted upon the auto. There are any number of reasons why the exception might have been produced.

2.

	Small Frogs	Large Frogs
Spots	7	8
No Spots	13	2

   Yes, there is a correlation. Most of the small frogs (13/20 = 65%) are not spotted, but most of the big frogs are spotted (8/10 = 80%). It appears that there is a greater chance of being spotted if you are a large frog.

3. Yes, most of the smokers died of lung cancer (5/7 = 71%) while most of the nonsmokers died of something else (5/6 = 83%).

4. a. Amount of income and level of blood pressure.

   b. high income and low income; high blood pressure and low blood pressure

   c. Inverse. As the income goes up, the blood pressure goes down.

   d. No. Only a correlation is asserted. If cause and effect is involved, it is most likely that low income causes daily problems that cause high blood pressure. However, it is possible that high blood pressure causes behavior problems that keep people from being productive and earning a high income.

CHAPTER 5

Pages 107–108

1. H < H, T   T < H, T   1/4 = 25%

2. 1/36 = 2.7%

3. a. 3/4 = 75%

   b. 9/16 = 56.25%

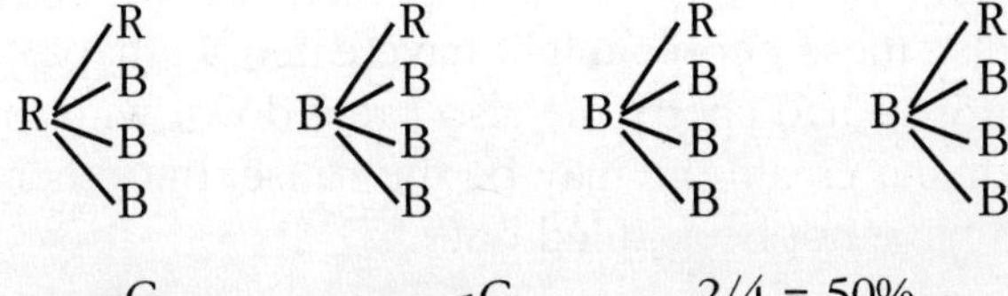

4. G < G, B   B < G, B   2/4 = 50%

5. 0 < 0 … 9   1/100 = 1%
   1
   2
   3
   4
   5
   6
   7
   8
   9 < … 9

Pages 112–114

1. 10/32 = 31.25% of the time this result will occur due to chance alone. Set up a tree diagram. The diagram will show 32 possible

results. Ten of the results will show three left turns and two right turns (e.g., LLLRR, LRRLL, LRLLR).

2. a. 1/2

   b. 1/2 = 50%

   c. If we ignore the people who could not tell the difference, we get 3 people who prefer Brand X and one who prefers Brand Y. A tree diagram analogous to the one for the four mealworms shows that there are 16 possible results. Out of these 16, four of them show three Brand X choices and one Brand Y choice (i.e., XXXY, XXYX, XYXX, YXXX). Therefore, the probability that this result is due to chance alone is 4/16 = 1/4 = 25%. If you concluded that the results indicate that people, in general, prefer Brand X to Brand Y, you stand a 25% chance of being wrong and a 75% chance of being right.

3. a. Although 4 of 5 = 80% is a high percentage, we do not know what else these people might have eaten. Perhaps 4 out of 5 people who had cherry pie also came down with the illness. The banana cream pie may be the cause, but other possible causes have not been ruled out.

   b. No, we need to know what percentage of the non-banana cream pie eaters got stomach pains and nausea.

4. Assume that a plate can contain from 1 to 7 letters or digits chosen from 26 letters and 10 digits, or 36 characters. Assume that any one or more of those characters can be blank, with the exception of all blanks (a plate must contain at least one letter or digit). Then there are $37^7 - 1$ possible plates.

5.

	Possibilities	
Door 1	Door 2	Door 3
(1) CAR	PIG	PIG
(2) PIG	CAR	PIG
(3) PIG	PIG	CAR

(1) If the car is behind door 1 (first option), and you switch to door 2, you lose.

(2) If the car is behind door 2 (second option), and you switch to door 2, you win.

(3) If the car is behind door 3 (third option), and we assume that the host, knowing what is behind each door, opens door 2 to show you a pig, and you switch to door 3, then you win.

Therefore, switching leads to winning in 2 out of 3 cases. If you do not switch, you win in only 1 out of 3 cases. You should switch!

6.

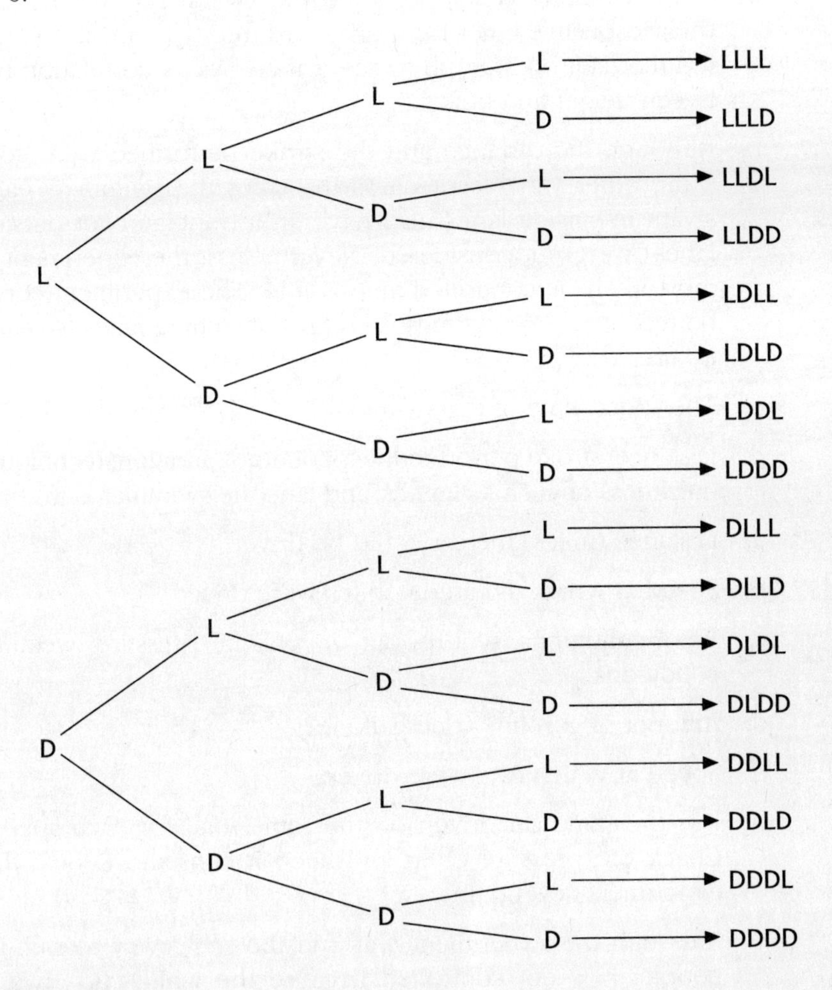

Pages 120–122

1. The height the balls bounced.
2. The height from which the balls were dropped; the surface on which the balls were dropped; how the balls were dropped.

3. The number of push-ups that could be performed.

4. Where the push-ups were performed (land vs. water)

5. Jars numbers 1 and 5, because all other variables in these two jars were kept constant (plant type, plant part, color of light). The only variable that was different was the temperature.

6. a. The swimming coach recorded the swimmers' heights and their respective times for the 50-yard freestyle. She then plotted the data on a graph to see if there was a correlation between height and time.

   b. Although the distance and the stroke performed were constant, other variables such as the lanes the swimmers each swam in (inside lanes may have an advantage over outside lanes) were not considered. Nevertheless, the experiment is most likely as controlled as possible. The experiment could be repeated several times each time switching lanes to make it fairer.

   c. The speed/time of the swimmers.

   d. Distance; stroke performed; experience, condition, technique, quickness of each swimmer, and lane the swimmer is in.

7. a. Distance (miles) the car could be driven.

   b. Speed at which the automobiles were driven.

   c. Road conditions, type of car driven, tire pressure, weather conditions.

   d. Amount of gasoline (one gallon).

   e. Speed at which cars were driven.

   f. Use the same car driven on the same road for each speed. Check tire pressure often and keep it constant. Repeat the experiment several times.

8. a. Although the experiment was run the same way for all 15 people, it is not controlled because the white cup always was first and orange second following the cracker. To be fair, the order would have to randomized and the cracker added before each or before none.

   b. Possibly because the white cup was given first when they were thirstier, or because the cracker spoiled the taste of the second cup, or because people prefer white (it looks cleaner or cooler, etc.), or perhaps the result was due to chance. The list of possibilities is endless.

c. No experiment can "prove" anything, because alternative explanations can always be found. This is especially true in this experiment.

d. Randomize the order. Eliminate the cracker. Repeat at least 100 times.

Pages 131–134

1. a. 24 mutations

   b. 6.4 mutations

2. 128 kph, 74 kph

3. 2.67 cups of milk and 3.33 eggs

4. 14 years old

5. $\frac{\sqrt{720 \text{ A}}}{\sqrt{370 \text{ B}}} = \frac{6 \text{ seconds}}{x \text{ seconds}}$ $x = 4.3$ seconds

6. The second store would save you about 1¢/pair. If you need the 8 pairs, shop at the second store. If you only need 6 pairs, shop at the first store and save $1.25.

7. Amount of oil and amount of gasoline; constant proportion; proportional relationship

8. Velocity and distance from each other; constant difference

9. Shadow length and time of day; proportional

10. 56 kilometers

11. 88.64 inches or 7.39 feet

12. 21 francs

13. $\frac{7°}{360°} = \frac{800 \text{ km}}{x}$ $x = 41{,}142.9$ km

CHAPTER 6

Pages 145–146

1. a. to establish the truth...

   b. Scientific evidence can not establish truth of hypotheses/theories (that is, explanations) in any ultimate sense because these explanations must be tested using *if ... and ... then ... therefore* reasoning coupled with observations of nature (experiments). When a particular explanation gains support by having its' predictions born out by experimental results, some other unimagined explanation may lead to the same predictions, thus also gain support. Conversely, when some explanation leads to predictions that are not found, the explanation has not been supported, but it has not been "disproved" because their may have been an unnoticed flaw in the experiment (that is, some uncontrolled variable). Because nature is complex and no experimenter can be certain that all variables have been accounted for, some doubt must always remain. Science can establish the "correctness" of its' hypotheses and theories "beyond a reasonable doubt" but not beyond any doubt.

2. a. to establish the correctness of some statement

   b. The dictionary definition fails to express important differences between the way scientists and mathematicians may use the word *prove*. The first thing to keep in mind is that mathematicians are not concerned with the real world with all of its complexities. Thus, for them proof involves starting with some initial statements that are assumed to be true (correct), and then goes on to use logical reasoning (deduction) to generate additional statements that follow and can also be assumed to be correct. For example, suppose that we have a set of cards each with a number on one side and a letter on the other. Suppose we assume that cards with vowels on one side have even numbers on the other. Then it follows by deduction that a card with an E on one side has an even number on the other. If the premises are true then the conclusion logically follows. This process of deduction is also used in science to generate predictions from the assumed truth of particular hypotheses and some imagined test conditions (for example, a controlled experiment). But in science the task is not yet complete. Scientists must now consult nature (do experiments) to see if the logically derived predictions do in fact occur. Only then are they able to reach conclusions. But their conclusions are not logically derived statements. Rather the conclusions are statements about the relative correctness or incorrectness of hypotheses/theories in light of the empirical results.

3. a. ability to walk, talk, ability to feed ones self, ability to put on and take off clothes

   b. ability to cry, ability to move limbs

4. a. earth, solar system, galaxy, universe

   b. The biosphere plus the hydrosphere (water), atmosphere (air), lithosphere (solid portion of the earth) together make up the planet called *Earth*. The Earth plus the other planets that orbit the sun (a star) together make up a *solar system*. All of the solar systems (we assume there are others) in the cluster of stars that contain the sun (the Milky Way) together make up a *galaxy*. All of the galaxies together make up the *universe*.

5. a. You think the painting is beautiful because the dog reminds you of a dog you owned and loved as a child. Your friend thinks the painting is ugly because it reminds him of a dog that bit him when he was a child.

   b. No. The question is one that can be answered at the organism level. Even though photons are most certainly involved in the detection of the painting by both individuals, our understanding of the situation in no way depends on that level of analysis.

6. a. the individual and/or population levels

   b. The most reasonable response (hypotheses) for the decline would center around factors that might be killing individual eagles or influencing their interactions with each other (for example, poisonous chemicals, loss of food/shelter due to forest cutting, loss of suitable mates).

CHAPTER 7

Pages 164–166

1. a. internal cell organelles

   b. leaves, roots, vascular tissues

2. a. yes

   b. The only way to scientifically test hypotheses, such as the hypothesis that an all powerful creator created the intermediate forms of life, is to use *if...and...then* reasoning to generate a prediction from the hypothesis. If the results of your test do not match the prediction, then the hypothesis has not been supported. Unfortunately, if the hypothesis asserts that an all powerful entity exists, then it follows that the entity can do anything it wants, hence lead to any predicted result. Hence their is no way to find a result that could not have been predicted by the hypothesis. Hence there is no way to find nonsupportive evidence. The hypothesis of an all powerful entity is untestable.

3. The control group represents the group used for comparison with the experimental (i.e., treatment group). Suppose, for example, we want to test the effect of student study groups on grades in a college biology course. The treatment then is to have students study in groups. The expected outcome might be that such students would achieve higher grades, but higher than whom? They should achieve grades higher than students who study alone—the control group. Hence, the control group results (that is, their grades) are compared with those of the treatment group to see if the treatment is effective.

4. a. Perhaps the salmon navigate by listening to familiar sounds that they heard and recalled from their trip to the ocean.

   b. To test this "sound" hypothesis, we would plug up the ears of some of the returning salmon. Such an experiment would be controlled if we were able to plug up the ears of the treatment group salmon in such a way that left them different from the control group salmon in only their ability to hear.

5. a. If the possible combinations of paired genes in the next generation were WW, WP, PW, and PP, and if one gene of the

pair dominates the other, then a 3 to 1 ratio results. But suppose instead, neither dominates. Then the WW offspring would turn out white; the PP offspring would turn out purple; and the WP and PW offspring would turn out some intermediate color such as pink. Hence we would have a 1:2:1 ratio of offspring types.

b. To explain a 1:1 ratio we would have to assume that the parents have only two of the three possible combinations of genes. For example, suppose that the combination MM does not exist. Therefore individuals could have only the MF or the FF combination of gene pairs. Suppose further that the MF combination is the male and the FF combination is the female. Then when a male produces gametes, the gametes can carry either the M or the F gene. When a female produces gametes, the gametes can carry either the F or the F gene. When these gametes combine, we get offspring with these possible combinations of genes: MF, MF, FF and FF. These combinations would produce a 1:1 ratio of males to females.

c. The more pairs of genes involved in producing a particular characteristic, the more intermediate forms there will be (e.g., 1:2:3:2:1 or 1:2:3:4:3:2:1 and so on).

6. Yes it does appear to contradict the postulate. But we need not necessarily reject the theory. After all, the result does not contradict the other postulates. Further, all we need to do to "save" postulate 6 is to slightly modify it to state that the factors *usually* assort independently.

7. If the genes are primarily responsible, and we raise genetically identical twins under vastly different environmental conditions, then their levels of intelligence should be similar. On the other hand, if the environment is primarily responsible, then their level of intelligence should be dissimilar. The converse situations could also test the hypotheses.

CHAPTER 8

Pages 178–179

1. Answers will vary. Here is one possible alternative hypothesis and a way to test it. Hypothesis: Nerve cells are highly active

during physical exertion. It is CO_2 produced by nerve cells that causes the CO_2 increase in the blood. Test: After exercise compare amount of increase in blood CO_2 in the blood near the brain (largely nerve cells) with the increase in the blood near muscles in a leg. If the hypothesis is correct, then the amount of CO_2 increase should be greater near the brain than in the leg.

2. If the center exists and functions as stated, and the center is destroyed surgically in a cat, then the proposed functions should not occur (i.e., when hydrogen ions are injected into the blood stream the heart beat rate should not increase).

3. Answers will vary. Here are three possible hypotheses: (1) The force of gravity causes the food to move down. (2) The phloem tubes and leaves form one continuous space. When food is created in the leaves, it collects and fills the spaces in the leaves. When more food is created, it pushes the old food down the tubes. (3) The phloem tubes are surrounded by tissue that is able to contract and squeeze the tubes. Because the tubes have one way values, the food is squeezed down but not up.

4. If hypothesis (1) is correct, and we turn a plant upside, then the food should not move. If hypothesis (2) is correct and we cut off the leaves, then the food in the top of the tubes should not move down. If hypothesis (3) is correct and we remove the surrounding tissue, then the food should not move.

CHAPTER 9

Pages 200–202

1. Strictly speaking the correct conclusion drawn was that evidence was obtained to support the hypothesis that the nucleus controls some of the activities in one specific type of cell. To generalize and claim that the nucleus controls all of the cell's activities is clearly unjustified. It certainly is not hard to imagine some activities that may be under the control of other cell parts or the cell environment. Further, just because the nucleus of this type of cell appears to control a specific activity, it does not mean that the nuclei of other types of cells act similarly. These possibilities would need to be tested with other types of activities and other types of cells before such generalizations are warranted. Over

generalization is a common mistake in the sciences and more so in everyday life.

2. Answers will vary. Here is one hypothesis: The spindle fibers are cork screw shaped. They attach to a chromosome in such a way so that when they begin to rotate, the chromosomes move in much the same way as a garage door attached to an automatic garage door opener moves.
3. Answers will vary. Experimentally testing the above hypothesis might be difficult. But if you could somehow grasp the spindle fiber with something that would prevent rotation, then the chromosome should not move. If you could magnify the fibers enough to get a good look at their structure, then they should appear banded.
4. No. At least not in any ultimate sense. See Chapter 6 question 1.
5. The cells would become over crowded with chromosomes, thus physically disrupting the activities of other cell components. Also with so many chromosomes producing so many chemical "directions" to other cell components, the other components would no doubt become unable to function correctly. This situation would be somewhat like a child with several sets of parents living in the house. If the over crowding did not disrupt the child's activities, having too many parents telling him what to do surely would.
6. The statement omits the effect of the environment. While it is correct to say that some characteristics are *largely* under the control of genes, none are solely controlled by genes. For any gene (i.e., segment of DNA) to express itself in terms of some observable characteristic, the gene's environment and that of the organism must at least provide minimal conditions for the gene to function.
7. This statement suffers from the same problem as did statement 6. All characteristics are controlled partly by the genes and partly by the environment. The relevant question is how much control is exerted by each.

CHAPTER 10

Pages 226–228

1. Answers will vary. One further experimental result that supports the atomic hypothesis is the so called "electrolysis" of water. An

electric current passed through water produces bubbles of gas near each of the two electrodes. When these gases are collected, they can be shown to be of two different kinds. The gas at one electrode causes a glowing piece of word to pop. The gas at the other electrode causes it to burst into flame. In addition, twice as much of the popping gas is produced. If it is the case that water consists of two kinds of atoms, then two kinds of gases should be produced. Further, the amounts of those gases should be in small whole number ratios (e.g, 2:1). Because both of these predicted results are found, the atomic hypothesis is supported.

2. No. Answers will vary. One piece of evidence that indicates that atoms can be split is the following. When tiny particles believed to be smaller than individual atoms are accelerated to very high speeds in large machines and directed at atoms placed in cloud-filled chambers, tracks of several different shapes appear that head off in several different directions. The interpretation of such a result is that the atoms have been busted into different kinds of "subatomic" particles that then produce the different kinds of tracks.

3. a. The large molecules come from the food and beverages that we consume.

   b. The small molecules are exhaled from the lungs, excreted in the urine or perspiration, or deposited as solid waste.

4. Assuming that you have some way to detect chemicals A, B, and C, one could first begin by placing samples of chemical A into three test tubes. Add enzyme 1 to the first, enzyme 2 to the second, and enzyme 3 to the third. Now test each test tube for the presence of chemicals A and B. Only the test tube that contains the enzyme that converts A to B will contain B. The others will contain A. Now place samples of chemical B into two test tubes and add the remaining two enzymes, one kind to each tube. Test each test tube for chemicals B and C. The tube that contains chemical C is the one that contained the enzyme that converts B to C. The remaining enzyme is the one that converts C to D.

5. a. Water would enter and the cells would burst. Water molecules will both exit and enter the cells. But more will enter than exit because there are simply more on the outside pushing in than inside pushing out. Hence the water molecules

will diffuse from areas of high water concentration (outside the cells) to areas of low water concentration (inside the cell).

b. Water would exit and the cells would shrivel up for the same reasons as stated above but here the starting water concentrations are reversed. So the net water movement will be reversed.

c. Cell size would remain the same because the net water movement into and out of the cell would be equal.

6. The forces are pushes due to collisions with other molecules. The movement is from high to low concentration because more outward pushes (from the high concentration area) will occur than inward pushes (from the low concentration area). The situation is somewhat similar to one in which a crowd of pushy people are moving from one end of a hallway to the other. If you get in their way, you will be pushed down the hall from the area in which they started (the high concentration area) to the other end of the hall (the low concentration area).

7. The other molecules typically are larger than the water molecules so they are not able to move through tiny pores in the cell membrane to escape. These larger molecules could in theory block some of the pores and prevent water from entering. Normally, however, more water molecules will move in simply because there are more on the outside, per unit volume, than on the inside; therefore, there will be more collisions with other water molecules that bump them into the cell, than out of the cell. This *net* movement into the cell will continue until some other force becomes greater and stops it. That other force could be the force of the cell membrane pushing in.

CHAPTER 11

Pages 260–263

1. The candy bar would be better because it consists largely of sugar molecules while the green salad consists largely of carbohydrate molecules. Fewer steps are involved in breaking down sugar molecules than carbohydrate molecules to release usable energy; therefore, for a quick source of energy the candy bar would be better. However a candy bar with chocolate in it would be a problem because chocolate contains fat molecules

that tend to slow digestion. Salad dressings also tend to contain fat molecules.

2. Less air, hence fewer oxygen molecules exist in Denver than at sea level. Therefore, sea level would produce faster times because oxygen molecules are needed to "burn" food molecules to release energy needed by the runners.

3. The initial materials and the end products are the same. The primary difference is the rate at which the molecules in the log and in the food are broken apart. The rate at which a log burns is controlled primarily by the supply of initial materials (i.e., log and oxygen). The rate at which food molecules are broken down is controlled primarily by a series of complex biological molecules (enzymes).

4. Answers will vary. See Chapter 10 Question 4 for one strategy. Another strategy that might work is to allow the reaction to proceed for varying lengths of time and then stopping it by some chemical or physical means. Then check to see what molecules are present. Those that are present after the short period of time has elapsed since the start of the reaction are presumably the first intermediate molecules to be formed. If the molecules could be tagged with some radioactive tag such as carbon 14 another strategy could be used. Tag one type of molecule in the sequence and let the process go to completion. Then determine the ratio of carbon 14 to carbon in the end products. The earlier tagged molecule is in the sequence, the greater will be the ratio of carbon 14 to carbon in the end products.

5. Presumably the fan would not turn because there are no nearby air molecules to heat up and bounce away from the blades which would cause them to turn.

6. If photosynthesis takes place outside the chloroplasts, then a plant cell that is given the proper raw materials and conditions for photosynthesis but lacks chloroplasts should be able to conduct photosynthesis and produce glucose and oxygen. On the other hand, if chloroplasts are necessary for photosynthesis, then glucose and oxygen should not be produced.

7. Two bands should show up as before. The top band should be the N^{14} band. The bottom band should contain the hybrid N^{14} and N^{15} band right where it appeared before. However this bottom

band should appear fainter than before because only half as many hybrid molecules should be produced in the second generation.

8. After four generations there should still be the same two bands but the amount of hybrid DNA should be less and less with each generation and the amount of N^{14} DNA should be more and more. After two generations only one fourth of the DNA strands should contain the N^{15} isotope. After three generations only one eighth of the strands should contain the N^{15} isotope and after four generations only one sixteenth of the strand should contain the N^{15} isotope. Hence with each new generation, the N^{15} band on the graph will get progressively lower while the N^{14} band will get progressively higher.
9. These other molecules must come from the food and beverages that we consume.
10. Answers will vary. Here are two hypotheses: Diffusion might cause them to move. There might be some sort of active transport as well.
11. Answers will vary. Here are two hypotheses: Perhaps the cell contains tiny structures that act as paths or streets for the movement of molecules. Perhaps the initial movement is random, but certain molecular forces exist to attract and hold the appropriate molecules when they move nearby.
12. Because there are twenty different amino acids the code must consist of more than one letter. Assuming that the order of the letters does not matter, a two letter code could code for only six different amino acids (e.g., AT, AG, AC, TG, TC, GC). Even if the letter order did matter, a two letter code could only code for 12 amino acids. Even if we assume that order does not matter, a three letter code could work because 24 possible words could be produced (e.g., ACT, ACG, ATG, ATC, AGT, AGC, GAT, GAC, GTA, GTC, GCT, GCA, etc.). In fact, evidence indicates that each three-letter code is a gene.

CHAPTER 12

Pages 292–298

1. The jar would be half full at 11:59 pm. When the jar is only half full it appears to have a lot of space left for growth. In reality,

however, the amount of time before all of the remaining space is gone is extremely limited. Likewise it may appear that there is considerable space and/or resources available for continued growth of the human population on Earth. Yet it is probably later than most people realize.

2. Answers will vary.

3. Line a probably represents the prey and line b represents the predator. Note that at around month 55, line b drops to near zero. This time is followed by a rapid increase in line a. It seems reasonable to interpret this as the predator population dropping to near zero (line b) followed by an increase in the prey population (line a) due to a lack of predation. It does not seem reasonable to think of it the other way around (i.e., a drop in the prey population to near zero followed by a rapid increase in the predator population).

4. a. True. A comparison of graph A with B tests the effect of temperature while a comparison of graph C with D tests the effect of initial number of adults.

   b. False. This is not a controlled comparison because not only does temperature vary, but so does the species of beetles and the initial number of adults. Therefore we cannot tell which of these factors is responsible for the outcome.

   c. False. We do have evidence that this particular adult population grows better at 20° C than at 35° C, but we should not generalize to all adult populations until others are tested.

   d. Possibly true. Two species have been tested and their largest adult populations appear to be at 40 individuals. But note that only two temperatures have been tested. Possibly other temperatures would produce larger adult populations. At this time we really do not know the optimal temperature.

   e. False. A comparison of graph C with D tests the effect of initial adult population size on number of larvae. Initial adult population size varied greatly but number of larvae did not.

5. Producers are needed to capture solar energy, thus they could not be eliminated. Decomposers are needed to recycle nutrients from dead organisms back into the substrate for use as raw

materials by the producers. Consumers could be eliminated as they neither capture energy or recycle nutrients.

6. Alternative (d) is the most likely cause. Without nutrients the plants would die followed by death of the primary consumers followed in turn by death of the higher-order consumers such as the bass. This process would take some time; hence, the bass deaths at the end of the summer. Alternative (a) is possible but it seems unlikely because of the time delay (i.e., had the poison killed the bass, it should have occurred right away). Alternatives (b) and (c) are not likely because bass do not use decomposers as a food source.

7. Sousa's observations of the red and green algae under dry sunning conditions lacked a control, thus we can not be certain that it was the dry/hot conditions that caused the observed effect. Other alternatives exist. Perhaps exposure to the air with its' increased amount of oxygen favored the red over the green algae. Perhaps the nitrogen in the air favored the red algae. To test these alternative hypotheses, controlled experiments would need to be set up in which the oxygen and then the nitrogen content of the air was varied to see what, if any, effect they would have on the two types of algae.

8. Perhaps the red algae has a coating of material on its surface that prevents water from escaping. To test this hypothesis one could conduct a controlled experiment comparing the survival rate of normal red algae with red algae with its external coating removed. If the hypothesis is correct, then the normal red algae should survive longer under hot/dry conditions. Perhaps the red algae survive better because they are able to absorb and store more water in their cells than the green algae. To test this hypothesis one would have to compare the water content of the two types of algae immediately after having been removed from the water. Dry the outside of the algae. Then weigh the algae. Then crush the samples and heat them to remove all of the internal water. Then weigh the dried samples. If the hypothesis is correct, then the percentage of weight loss should be greater for the red algae than for the green algae.

9. The existence of such a wide range of climax communities in Arizona can be explained by the fact that a considerable range in elevations occur across the state. Consequently air temperatures

and rainfall varies considerably. Cooler temperatures and increased rainfall produce different climax communities because different types of plants and animals have adapted to survive under these different conditions. Chapter 13 will discuss how such adaptations are acquired.

CHAPTER 13

Pages 318–321

1. In a sense the theory of biogenesis and the present theory of the origin of life on Earth are contradictory. The way that most biologists resolve this possible contradiction is to argue that the physical conditions that existed on Earth billions of years ago, which presumably allowed for the spontaneous generation of complex organic molecules and eventually life, were vastly different than present conditions. Much evidence has been obtained to support this view.

2. See Chapter 7 Question 2. Such theories are normally considered to be outside the realm of science because they contain claims that can become untestable. Science deals only with explanations that can be tested either with available techniques or at least are explanations that have the potential to be tested with new techniques. No techniques exist for the testing of explanations that invoke supernatural powers because such powers, by definition, can do anything, hence can lead to any and all possible predicted results.

3. Intermediate forms support the theory of evolution and do not support the Genesis theory of special creation. Evolution theory predicts that intermediates should exist as these represent forms that are in the process of change of one form to another. Special creation theory predicts no intermediates because it claims that each kind of organism was created during a brief period of time "each according to their own kind."

4. Strictly speaking, the large number of organisms does not support or contradict either theory as neither theory stipulates the number of species that should exist. However, to most observers, the vast numbers of species seems more understandable within

the theory of evolution than within the theory of special creation. There exists tremendous variation in environmental conditions on Earth, hence it seems reasonable that organisms would have evolved to exist under these varied conditions. Likewise it seems unreasonable to think of a creator, albeit a very powerful one, taking the time to create 35,000 different kinds of fungi!

5. No. Any piece of evidence that supports evolution theory could also have been the work of an all powerful creator. The creator, by definition, is powerful enough to produce any result, hence no result could contradict the postulation of his/her existence. Such postulates can not be scientifically tested (see Chapter 7 Question 2).

6. The numbers do reveal a correlation. The probability of finding a dark moth is much less in the light woodlands (3/327 = 1%) than it is in the dark woodlands (1403/1547 = 91%). The results support Darwin's theory of natural selection because that theory would lead one to predict that dark moths would be more abundant in dark environments and vice versa. Of course other theories would make the same prediction.

7. Answers will vary. One piece of evidence that supports postulate 1 is that body builders lift weights and acquire large muscles. Likewise when muscles are not used, as is the case when a broken leg is placed in a cast, the leg muscles become smaller and weaken. To test postulate 2 one could cut off the tails of a group of adult mice. If postulate 2 is correct, then the mice should produce offspring without tails.

8. Natural selection can act on the available variations that exist in populations; but it cannot create new variations. Mutations are important because they are the source of new variations. Without new variations to be selected for or against by natural selection, the process of evolution would not get far.

CHAPTER 6 SAMPLE EXAM

1. d. Flasks 1 and 2 vary in only one way (i.e., plugged vs. unplugged); therefore, a comparison of 1 and 2 will tell whether or not the plug makes a difference. Flasks 1 and 5 differ in only one way (i.e., untreated versus sterilized). Other comparisons involve more than one difference; therefore, they can not serve as controls.

2. a. Flask 5 is most open to the air; therefore, it has a greatest chance of having microbes or their spores fall in.

3. c. Spallanzani's and Pasteur's experiments made it clear that spontaneous generation does not occur, thus no microbes should appear in flask 6 because those initially present were presumably killed by sterilization and no new ones can enter from the air.

4. a. Presumably the only difference of any consequence between flasks 5 and 7 is the size of the opening; therefore, it is the likely cause of the difference in cloudiness. Presumably a larger opening increases the likelihood of microbes falling in. The growth of microbes produces the cloudiness.

5. d. The biogenesis hypothesis leads to the prediction that no microbes should appear in flasks 6 and 8 because all old ones were killed by sterilization and no new ones can enter from the air. Thus, if this result were found, it would support the hypothesis.

6. d. The vitalistic or abiogenesis hypothesis would lead one to predict the growth of microbes in flasks 6 and 8, whereas the biogenesis hypothesis leads to the opposite prediction (see 5 above). Therefore, what happens in flasks 6 and 8 is crucial to the test of both hypotheses.

7. c. Flasks 5 and 7 are both open to the air; therefore, the hypothesis under question would predict growth in these two flasks. Growth could occur in flask 1 due to lack of sterilization. Growth in flask 6 would not support the hypothesis because the hypothesis predicts no growth.

8. a. Clearly Spallanzani was an advocate of the biogenesis hypothesis. This hypothesis predicts no growth in flask 8 because old microbes were killed and no new ones can fall in.

CHAPTER 7 SAMPLE EXAM

1. d. The biological classification system is a hierarchy that first divides all organisms into large groups called kingdoms. The kingdoms are then subdivided into phylums, classes, orders, families, genera, and species categories. Each subsequent category includes fewer organisms that share more characteristics with others in that category; therefore, they are more closely related in terms of their ancestry. The smallest mentioned is genus (plural:genera).
2. b. I and III are the organisms that are in the same genus (Archips); therefore, they are the most closely related.
3. b. Organisms I and III must be in the same family (Tortricidae) because they are in the same genus (Archips). Therefore, the only organism that is not in the Tortricidae family is organism II. Therefore organism II must be the most distantly related to organism IV.
4. e. Organism I is in the class Hexopoda as stated. Organism II must also be in the class Hexapoda because it is in the order Lepidoptera, which is in the class Hexopoda (see column 1). Organism III must also be in the class Hexapoda because it is in the genus Archips, which is in the class Hexapoda (see column 1). Organism IV must also be in the class Hexapoda because it is in the family Tortricidae, which is in the class Hexapoda (see column 1).
5. b. Choice c is possible but not as likely as an explanation that assumes that white hair is caused by recessive gene and black hair is caused by a dominant gene. If this were the case, then the parents each had to carry at least one dominant gene for black hair. Did both carry a recessive gene for white hair? They must have, because, if they did not, each offspring would have carried at least one dominant gene and none of the offspring could have been white. One was white; therefore, both parents had to carry both a dominant and a recessive gene.
6. c. An Aa individual can produce gametes with the A or the a gene. An aa individual can produce gametes with an a gene or the a gene (i.e., both are the same). All possible combinations of these gametes are Aa, Aa, aa, aa. Note that his will produce 2 of 4 or 50% Aa individuals and 2 of 4 or 50% aa individuals. Therefore, 50% of the offspring will have the recessive aa genotype and phenotype.

7. c. A Ddee parent can produce four possible gametes as follows: De, De, de, de. Of these four, two show the de genotype, therefore the probability is 2/4=50%.

8. b. First generate the possible gametes:

   DdEd = De, de, dE, de for one parent.

   DdEE = DE, DE, dE, dE for the other parent.

   Now cross the gametes:

	DE	DE	dE	dE
DE	DDEE	DDEE	DdEE	DdEE
De	DDEe	DDEe	DdEe	DdEe
dE	DdEE	DdEE	ddEE	ddEE
de	DdEe	DdEe	ddEe	ddEe

   There are 16 possible offspring genotypes, and four of them have the DdEe genotype, 4/16 = 25%.

CHAPTER 8 SAMPLE EXAM

1. c. The graph shows that as the percentage of oxygen in the arterial blood (one variable) goes up, the rate of blood flow in the dog's leg (the other variable) goes down. Thus, we have two variables that covary, that is, are correlated. Thus, we have a statement of the results. Note that we have no proposed explanation (i.e., hypothesis) for them as yet.

2. b. This sentence proposes a possible explanation (a hypothesis); however, the hypothesis is contradicted by the results because it leads one to predict that as the % oxygen goes down so would the rate of blood flow. But the opposite result was found.

3. a. This sentence proposes a possible explanation (a hypothesis) that argues that oxygen lack causes an increase in blood vessel size that would increase rate of blood flow (the prediction). Because this predicted result was found, the hypothesis that leads to the predicted result is consistent with the results.

4. b. This sentence proposes a possible explanation (a hypothesis); however, the hypothesis is inconsistent with the results because

it leads one to expect (predict) that as the percentage of oxygen in the blood goes up, so would the rate of blood flow. But this was not found.

5. c.

6. a. This is a much more direct test of hypothesis I:

   If... nerves stimulate the pancreas to secrete its enzymes into the intestine...

   and... the nerve to the pancreas is stimulated...

   then... the pancreas should secrete enzymes.

   Result... The pancreas did secrete enzymes.

   Therefore... the hypothesis has been supported.

7. b. Strictly speaking, this result contradicts hypothesis I.

   If... nerves stimulate the pancreas to secrete its enzymes into the intestine...

   and... the nerve to the pancreas is cut...

   then... the pancreas should not be able to secrete enzymes.

   Result... But the pancreas did secrete enzymes.

   Therefore... the hypothesis has been contradicted.

If we assume that the pancreas is stimulated by either nerves or hormones and it is not nerves, then it must be hormones (i.e., support for hypothesis II).

8. b. The primary functions of the digestive tract are to break up large pieces of food and break up large food molecules into smaller ones and then have these small molecules absorbed into the blood stream. An animal that uses blood as its food would not need to break down its food much. However, the animal would still need some digestive tract, primarily for the absorption of the blood/food into its own blood stream.

9. c. Answer a is possible but not likely because most of the chemicals enter the stomach and the food is already broken down to a considerable extent by the time it reaches the small intestine.

Answer b is not likely because soft, folded tissue will not offer much, if any, support. While it is true that the small intestine connects the stomach to the large intestine, if this were its only function it would not need folds and a rich blood supply. Therefore, answer d is not likely to be correct. Answer e is also not likely as most signals from the brain are conducted via nerves and do not require many folds and a rich blood supply. Answer c is the best choice, as the folds would increase surface area for absorption and the rich blood supply would be used in carrying the absorbed food to other parts of the body.

CHAPTER 9 SAMPLE EXAM

1. d. A cell that travels around the body would need to be fairly streamlined, hence, cell b or d is a possibility. Cell d is probably better suited for travel in tubes. Cells shaped like b could get jammed up like logs in a log jam. Therefore, d is the best choice.

2. a. A cell that picks up and transmits signals would need to be long like an electrical cord. Therefore, cells a and b are possibilities. Cell a is the better choice because it appears to have lot of projections at one end that could possibly pick up signals from a number of cells and a few projections at the other end to pass a common signal to one or two other cells.

3. b. The shape of cells a, b, and c make them all possible candidates. Cell a has already been used, so that leaves cell b or c. Cell c is probably not as well suited for lengthwise contracting as cell b; that is, What would happen to its large nucleus? Also, why does cell c have little projections on the top and not on the bottom? Cell b appears to be the best guess at this time.

4. e. The choices left are cells c and e. Both could be found in skin. Therefore, let's look at item 5. Note that item 5 asks for a cell capable of absorbing food molecules. Cell c is the better choice for this because of its tiny projections at the top that may be used to increase surface area for absorbing food. This leaves cell e as the best choice for item 4. Note that its projections appear to be well suited for expansion and contraction.

5. c

6. b

If... the cells are three different types of cells each with its own specific characteristics and appearance...

and... we observe living cells over an extended period of time...

then... they should not change their appearance. (This last statement is the prediction, answer choice b.)

Answer choices a, c, and e are neither included nor excluded by this hypothesis, hence are irrelevant. Choice d clearly is not predicted by the hypothesis.

7. d. Choice d is not predicted by the first student's hypothesis. Instead it is predicted by the second student's alternative hypothesis:

    If... the cells that differ in appearance do so because they have been killed and preserved at different points in the division process...

    and... we observe living cells over an extended period of time...

    then... we should see a change in appearance of one type of cell into another type (i.e., what we would really be seeing is one type of cell going through changes in appearance, answer choice d).

8. b. All of the answer choices are reasonable hypotheses to account for the greater amount of yolk in birds' eggs. However, the only one that is consistent with the evidence is choice d. Mammals depend on continued nourishment from the female parent for development, typically via the placenta.

9. d.

    If... petals produce a chemical necessary for fruit development...

    and... the petals are removed from the flowers on one plant but not another...

    then... fruit should develop only from the flowers with petals.

    This last statement is the prediction that logically follows, answer choice d.

10. e. The percentage of developed fruit on the plant with the petals removed is 6/20 = 30%. The percentage of developed fruit on

the plant with petals is 10/30 = 33%. These percentages are about the same; therefore, the petal hypothesis has not been supported. To have supported the hypothesis, the second percentage would have to have been much higher than the first.

11. b. There is an inverse proportional relationship between magnification and the number of yeast cells you can see: the higher the magnification, the fewer cells seen. At 15X magnification (lower power) 100 cells can be seen along the diameter; therefore, at 45X (higher power), you can see only 15/45 of the cells because it is 45/15 times the magnification.

12. c. A 10X eyepiece and an objective lens of 15X will magnify 10X × 15X = 150X.

CHAPTER 10 SAMPLE EXAM

1. a. The positive iodine test provides evidence that starch molecules are present. The positive Benedict's test provides evidence that sugar molecules are present. We know that starch molecules consist of linked sugar molecules; therefore, the statement is probably true.

2. c. The iodine test for starch was not run; therefore, we have insufficient data to conclude that unknown six contains starch molecules.

3. b. Potatoes contain starch, so if unknown 7 was a potato, it should have tested positive for starch (the iodine test). It did not; therefore, unknown 7 is probably not a potato.

4. c. Let us assume that the molecules in either the liver or in the hydrogen peroxide (H_2O_2) or in both change when bubbles are produced (i.e., a gas of some sort is produced).

If...	the liver molecules but not the H_2O_2 molecules change...
and...	new liver and new H_2O_2 are mixed...
then...	bubbles should be produced in tube 1;
and...	old liver and new H_2O_2 are mixed...
then...	bubbles should not be produced in tube 2;

and...	old liver and old H_2O_2 are mixed (tube 3)...
then...	bubbles should not be produced in tube 3;
and...	new liver and old H_2O_2 are mixed...
then...	bubbles should be produced in tube 4.

Because only this last prediction is listed as an answer, option c must be correct.

5. b. The actual result listed in the table for tube 4 is that bubbles were not produced. Because the hypothesis led to the prediction that bubbles should have been produced in tube 4, the hypothesis has been contradicted.

6. b.

If...	liver molecules do not change, and H_2O_2 molecules change when mixed...
and...	old liver molecules and new H_2O_2 molecules are mixed...
then...	bubbles should be produced.

This is the experiment and predicted result for tube 2, answer choice b. None of the other answer choices follow from the hypothesis.

7. a. If the cells burst, then they must have taken in water (or some similar liquid). Blood cells contain some salt and some water. For the water to have entered the cell, the concentration of water molecules must have been higher on the outside of the cell than on the inside. Recall that molecules tend to diffuse from areas of high to low concentration. Answer choice a (used distilled water) would therefore produce a movement of water into the cells and a possible bursting of their membranes. Adding very salty water would produce the opposite effect. The speed at which the water is added probably makes no difference as would whether the cells are alive or dead, because the phenomenon is a purely physical one.

8. b. The cell is in a 10% salt solution; therefore, water molecules are more highly concentrated on the inside than on the outside. Consequently, water will diffuse out and the cell will shrivel.

9. a. See comment for item 8. The water concentration is higher on the inside, because the water is diluted by the salt more on the outside.

10. a. Diffusion is a physical process. The greater the surface area/volume ratio, the faster diffusion through the membrane will take place. Therefore, given cells of equal shape, the smallest cell will be the most efficient. In other words, small cells have a greater surface area for the given volume than do large cells.

11. e. See comment for 10.

12. d. Given 10 ml of the 5X solution, you would need to have 5 times as much freshwater to reduce the concentration to 1/5 of what it is (i.e., to produce a 1X solution, the concentration of normal sea water). Five times 10 ml of water is 50 ml. Therefore, you would need a total of 50 ml of water to produce a solution of normal sea water. You already have 10 ml, so add 40 ml.

13. d. Assuming that we now have a 1X solution (from item 12), to produce a 1/2X solution we need 2 times as much water. Therefore, instead of 50 ml, we need 2 × 50 ml or 100 ml of water. We already have 10 ml so add 90 ml.

CHAPTER 11 SAMPLE EXAM

1. a. The primary function of mitochondria is to break down food molecules to yield usable energy for the cells' activities. The fewer mitochondria, the less energy a cell is able to provide; therefore it can be inferred that a cell with only one mitochondrion must have low energy needs.

2. b. Answer a is possible but not likely. The evidence suggests that the endoplasmic reticulum (ER) deals with radioactive materials but not that it deals only with radioactive materials. Answer c can be eliminated because it is not as good a choice as b or d. The evidence could support either b or d, that is, the evidence supports the hypothesis that the ER is active in processing or transporting radioactive materials, which presumably are harmful. But answer b is a better choice than d, because we really have no evidence at this point that the ER can break down the possibly harmful radioactive molecules.

3. b. Variables that may influence the outcome of an experiment and whose values may vary independently of one another are called independent variables. Solution type, amount of light, temperature, and color of BTB at the start all fit this definition; therefore, they are independent variables.

4. c. The measured outcome, or outcomes, of an experiment is referred to as the outcome or dependent variable. Presumably, its values depend on one or more independent variables. The color of the BTB solution at the end of the experiment presumably may depend on solution type, amount of light, and so forth; therefore, it is the experiment's dependent variable.

5. d. To test the possible effect of the amount of light variable, we must choose test tubes in which the amount of light varies from one tube to another but the values of the possible effects (other independent variables) are the same. We do not want to confuse the issue; that is, we want to be certain that any difference that shows up is due to variation in the amount of light and not to something else. Tubes 4 and 5 are the only tubes that vary the amount of light (i.e., light and dark) and keep the values of the other independent variables constant (i.e., both have 5% sucrose, are at 10^0, and were blue at start).

6. e. CO_2 changes a blue BTB solution to green and then to yellow. If yeast produce CO_2, then the blue solution in test tube 2 would turn green or yellow, depending upon how much CO_2 was produced.

7. d. Notice that solution type, amount of light, and temperature all vary between test tubes 5 and 6. We know something caused the solutions to differ in color at the end of the experiment. However, because the values of three independent variables differed between tubes 5 and 6, we have no way of knowing which of these variables, or possibly which combination of these variables, was the cause (i.e., the experiment was not controlled). Also, note that yeast were present in both tubes 5 and 6, thus yeast alone could not be the cause.

8. a.

   If... virus DNA tagged with P* enters the bacteria...

   then... P* should be found inside the bacteria (choice a).

9. c.

If...	P* is found outside the bacterial cells...
then...	virus DNA probably did not enter the bacteria, (Recall that the P* is attached to the DNA so its presence indicates where the virus DNA is.)

10. b. S* is attached to the outer protein coat; therefore, its presence inside the bacteria would indicate that the outer coat had entered the cell (choice "b").

11. d. Results III and IV (answer choice d) indicate that the P* and S* have remained on the outside; that is, none of the viruses had entered the bacterial cell.

12. c. The best choice is probably c. Although it is true that the data provide evidence for the hypothesis that DNA consists of nucleotides, a component of nucleic acids, they are silent on the presence of deoxyribose and phosphoric acid. Therefore c is a better answer than a. The best answer might be that the data provide evidence for some parts of the statement, but this is not an answer option.

13. a.

If...	there is a specific pairing of nucleotides...
and...	the relative quantities of nucleotides are compared in various organisms...
then...	the relative quantities of specific pairs should be the same.
Result:	The relative quantities of specific pairs of nucleotides (e.g., adenine and thymine) are very close to the same (i.e., $1.13 \approx 1.11$, $1.15 \approx 1.14$, $0.84 \approx 0.80$)...
Therefore...	the data provide evidence for this statement (answer a).

14. c. The data tell us nothing about purine-pyrimidine pairings thus can provide no evidence about their possible role in keeping the diameter of the DNA molecules constant.

15. b. Note that the ratio of adenine to guanine is fairly constant for four of the five sources (calf, rat, virus, and sperm of rat) at about 1.2 to 0.9. However, the moth ratio of 0.84 to 1.22 is vastly different, therefore, the data do not provide evidence to support the statement that the adenine to guanine ratio is fairly constant for all species.

16. a.

If...	photosynthesis involves a reaction controlled by light...
and...	the amount of light is varied...
then...	the rate of photosynthesis should vary (prediction 1).
If...	photosynthesis involves a reaction controlled by enzymes (which are sensitive to variation in temperature)...
and...	the temperature is varied...
then...	the rate of photosynthesis should vary (prediction 2).

Answer choice a is the only one that embodies both of these predictions.

17. c. Bromthymol blue is a chemical that changes color from blue to yellow when it reacts with CO_2 molecules. Thus a change of water color from blue to yellow shows an increase in CO_2. The CO_2 increase presumably came from the fish's respiration.

18. d. *Elodea* is a green plant capable of conducting photosynthesis when light is present. Photosynthesis utilizes CO_2 and produces O_2, thus would counteract the increase in CO_2.

19. b.

If...	chlorophyll molecules are necessary for photosynthesis and photosynthesis produces starch...
and...	plants without chlorophyll molecules and no other obvious energy source (i.e., albino plants with no cotyledons) are grown...
then...	they should not be able to conduct photosynthesis and produce starch. (This is the prediction that follows, i.e., answer choice b).

CHAPTER 12 SAMPLE EXAM

1. e. The aquarium was sealed, so presumably nothing can get in or out. Because nothing can get in or out, the weight will not change. This would be the case regardless of the condition of the organisms inside.
2. c
3. a
4. a
5. b
6. e
7. d

Statements 2–7 can be organized as follows:

Observation: Several fly maggots can be seen crawling in and on both fish (item 2).

Causal question: Why is the bluegill more decomposed than the bass (item 7):

Alternative hypotheses: The bass might have tougher skin (item 3). The difference in decomposition may be due to the amount of available moisture (item 4).

If... the difference in decomposition is due to the amount of moisture...

and... a dead bluegill and bass are placed within 1 foot of the lake (presumably making the amount of moisture the same...

then... they would decompose at a similar rate. This last statement, which is part of item 5, is a prediction.

Lastly, item 6 is a conclusion (i.e., we can't say for sure why they are decomposing at different rates) because it states the extent to which the hypotheses have been supported or not supported, or, in this case, why they are not adequately tested.

8. b. Although all five statements represent hypotheses, that is, possible explanations for why the flowers in the bed with the grass clippings were greener than those without, b is the most reasonable hypothesis because it is the only one consistent with

evidence; that is, we know that grass molecules can be broken down by decomposers and absorbed by roots. We have no evidence that it was wetter in bed 1 than in bed 2 (choice a). There is no evidence that flowers can eat grass clippings (choice c). There is no evidence that roots can absorb grass (choice d). Finally, there is no evidence that absorbing yellow substances produces a green color in flowers.

9. e. The two flower beds were treated in at least two different ways (i.e., one got grass clippings the other did not, one is on one side of the house, the other is on another side of the house, where it may be wetter, cooler, and so on). Thus, the two beds differ in more ways than one, so it is not possible to determine the cause (i.e., it is not a controlled experiment).

10. a. The wheat is the producer organism in this short food chain. The cow is the primary consumer and the people are primary or secondary consumers or both, depending on whether they eat the wheat, the cow, or both. Only about 10% of the energy consumed at any level of a food chain is stored; therefore, it is available to the next higher level. About 90% of the energy at each level is used to keep the organisms alive. Therefore, the people could get the most energy by eating the cow first before the cow eats the wheat and uses up 90% of the energy it could have provided them.

11. c. As mentioned in number 10, energy is lost as food passes to higher trophic levels. Less energy means less weight/mass can be maintained. Therefore, the tertiary consumers would have the least mass.

12. e. Carbon, oxygen, water, and phosphorous are all atoms or molecules, that is, all types of matter. Matter recycles. Energy, however, is not matter. It can be thought of as motion or potential motion of that matter. This motion "enters" the matter of ecosystems when moving photons of light shine on green plants. The plants "capture" the motion and then use, and lose, some of it to their surroundings; therefore, energy does not recycle.

13. c. There are two categories of tree characteristics listed in the table, Douglas fir is one category and cedar and hemlock is the other; that is, cedars' characteristics are not differentiated from

hemlock, therefore answers b and d can not be correct. Douglas fir seedlings would not grow in the dense forest, because the table tells us that their seedlings die in shade. Thus c is the correct answer. Note that the table tells us that cedar and hemlock seedlings grow in shade.

14. a. The fire would decrease tree density, therefore decrease shade and increase ashes. This would favor growth of Douglas fir seedlings. Note also that their seeds are winged. This would help them colonize a new area as well.

15. b. The diagram represents a series of events through time in which a larger lake/pond has been gradually reduced in size by sedge that appears to change to sedge peat and then to woody peat as it decomposes. Note that the plants able to grow on these substances change from sedge to shrubs to trees. This process is called succession.

16. c. Decomposition appears to change sedge peat to woody peat. Position III now appears to be sedge peat, so it would most likely become woody peat next.

17. Each graph displays several data points on graphs in which the increasing values of two unknown variables are plotted. The two variables in item 16 are length and weight of round brass bars all of the same thickness. In other words, we have a sample of bars that increase in length and weight but have the same shape, thickness, and type of metal. Because these other variables are held constant, as the bars get larger, they will get proportionately heavier, as depicted in graph B.

18. The two variables here are age and time to run one mile. In general, we can expect older people to take longer—as age goes up, so does time. This is a direct correlation. Note that unlike the bar situation other variables are not held constant. Therefore, some exceptions will occur and the pattern will look more like graph D.

19. The two variables are time and population size. The larger the time, the larger the population. Note, however, that the rate of increase will increase; for example, 2 flies could produce 4 offspring, these could produce 8, then 16, then 32, and so on. Therefore, graph A is the correct choice.

20. The two variables are weight and height. Generally, taller people weigh more but some exceptions occur, therefore, graph D is the best choice.

21. In this case, the graph is of a single variable—shell width—plotted against its frequency. Because the snails are all of the same species we can expect that snails will vary in width generally, with those of intermediate widths to be the most common; therefore, graph C is the best choice.

22. Here we have the values of one variable—salt concentration—plotted against the frequency of hatching. Brine shrimp are saltwater organisms, and their eggs can be expected to be influenced by osmosis. Therefore, eggs in freshwater will probably take in water via osmosis and perhaps burst. Eggs in very high concentrations of salt may lose water via osmosis and shrivel. The best hatching will probably take place at or near the middle of the range, that is, in normal sea water. Therefore, graph C is the best choice.

CHAPTER 13 SAMPLE EXAM

1. a. Without reproduction, species would become extinct, life would not continue, and no new species would arise. However, whether an individual reproduces or not has little, if any, effect on its own survival.

2. a. Sexual and asexual reproduction alike can provide for a consistency of traits and a continuation of species; therefore, choices b and c are not correct. Choice d is also not correct. With asexual reproduction, no eggs are fertilized. Hence, the answer must be a. Indeed, sexual reproduction allows genes to mix; therefore, it will produce a greater variety of organisms than asexual reproduction.

3. b. Choice a is not correct because evidence suggests that flowering plants are highly evolved. Choice c would result in less genetic diversity, not more, because there would be no mixing of genes from two separate individuals. Choice d is not correct because environmental influences tend to decrease genetic diversity by eliminating some individuals and their genes. Thus we are left with choice b and e. Of these, choice b is probably

better, because it is easy to imagine that the pollen and ova may not mature at the same time, while it is difficult to imagine why evolution would favor such flowers in changing environments.

4. d. The statement is not a question (no question mark). The statement is not data. The data are the actual fossils and the specific rock layers in which they are found. The statement is not a prediction. Predictions follow from hypotheses and experiments, neither of which are mentioned. The statement is an interpretation of the data. The interpretation (explanation or hypothesis) serves to explain the data. In other words, if..."animal groups tend to develop many kinds after they originate," and...we look at the fossil record, then...we would expect to see one kind of animal in lower rock layers and many kinds of similar groups in higher rock layers.

5. d. The diagram shows birds coming into existence about 170 million years ago. The next most recent group is the mammals, at 240 million years.

6. c. There are three groups of fish. The armored fish have become extinct. The number of species present today is reflected by the width of the top band in the diagram. The band for the bony fish is wider than that of the cartilaginous fish; therefore, the bony fish are the most successful.

7. c.

If...	species do not change through time (hypothesis)...
and...	the fossil record reveals species that have lived in the past...
then...	the same fossils should be found in old and new rock layers.

This last statement is the predicted result (answer choice c).

8. d. The problem can be set up as a proportion as follows:

$$\frac{1 \text{ inch}}{500 \text{ year}} = \frac{x \text{ inches}}{2{,}000{,}000{,}000 \text{ years}}$$

Solving for x will give the number of inches in the time line = 4,000,000 inches.

Since there are $1 \times 12 \times 3 \times 1760 = 63{,}360$ inches in one mile,

$$\frac{63{,}360}{1 \text{ mile}} = \frac{4{,}000{,}000 \text{ inches}}{x \text{ miles}}$$

Solving for x gives us 63.13 miles.

9. c. The original genotype is assumed to be the heterozygous DL genotype. If we cross DL with DL we get D or L gametes that combine to form DD, DL, DL, LL zygotes. The first three of these would all be dark because they carry the dominant D gene. The last one would be light. This is a three dark: one light ratio. This ratio would stay about the same after two or three generations.
10. b
11. c
12. a
13. d
14. a
15. b The causal question raised is why did the relative number of dark and light insects change; that is, why are there fewer dark insects and more light insects now? Statements listed in 10, 12, 14, and 15 are all possible explanations, hence they are hypotheses. Statement 10 is contradicted by the data that showed the initial ratio of dark to light to be 3:1 (see item 9). Therefore, the correct answer for 10 is b. Hypotheses stated in 12, 14, and 15 are not contradicted by data; therefore, must be considered reasonable. However, 12 and 14 are more probable hypotheses than 15 because they involve natural selection, a known process, but 15 requires white sand to cause dark insects to become light, and there is no evidence that such a process happens. Therefore, the correct answer for 12 and 14 is a and the correct answer for 15 is b. Statement 11 is a restatement of the data. It does not attempt to explain the data; therefore, is not a hypothesis. Finally, statement 13 states what is likely to happen to the frequency of D and L genes, assuming that natural selection continues to occur; therefore, it is a prediction.

INDEX

Abduction, 42, 47, 64–65
Abductive-deductive thinking, 43
Abiotic factors, 281
Acetabularia, 185–187
Acetic acid, 238
Acid, 220
Activation energy, 220
Active transport, 225
Adaptive radiation, 309
Adenosine diphosphate (ADP), 237
Adenosine triphosphate (ATP), 237
Alleles, 160, 162
Alternative hypotheses, 24, 42, 83
Amino acids, 218
Animals, 153
Anthropomorphism, 29
Applied science, 43
Artery, 172
Artificial selection, 322
Assumptions, 44, 62
Atoms, 136, 141, 210, 213

Base, 220
Base pairs, 252
Behavior, theory of, 157–158
Beliefs, 136
Biodegradable, 283
Biogenesis, 138, 184
Biological community, 142, 281
Biological magnification, 283–284
Biology, skin-out, 152
Biology, skin-in, 152
Biomass, 288
Biosphere, 142
Biotic potential, 272, 313
Blastula, 200
Blood circulation, theory of, 172
Calvin cycle, 249
Carbohydrates, 217
Carbon, 211
Cardio-accelerator center, 174–175
Cardio-inhibitory center, 174–175
Carnivore, 281
Carrying capacity, 273, 276–277
Catalyst, 220
Causal questions, 23, 63–64, 83
Causality, 106
Cell nucleus, 185–186
Cell theory, 184
Cell membrane, 225
Cell replication, theory of, 189
Cells, 141
Cellular fermentation and respiration, theory of, 242
Cellular respiration, 236–239
Centrifuge, 243, 254
Centripetal force, 210
Centromere, 190
Chance, 106
Characteristic, 152, 158
Chemical energy, 237
Chemical indicators, 229, 265
Chemical reaction, 212
Chemical bonds, 211
Chloroplasts, 246–247
CHNOPS, 22, 139, 216
Chromosomes, homologous, 191–192
Chromosomes, movement of, 189–190
Chromosomes, 162, 196
Chromosomes, replication of, 188–189
Citric acid, 238
Class inclusion, 60

Classes, 153
Classification, theory of, 156
Climax community, 287
Combinatorial thinking, 61, 65
Community respiration, 288
Competition, 274, 276
Concept, 40
Conclusion, 35–36, 67, 83, 109–112
Conservation reasoning, 60
Constant, 89
Consumers, first and second-order, 281
Control group, 68
Controlled experiment, 66, 116–118
Controlling variables, 62, 66–67, 115–120
Correlation, 89, 123
Correlational thinking, 62, 67–70, 88–101
Crossing over, 191–193
Cytokinesis, 188

DDT, 283
Decomposer, 282
Deduction, 35
Density-dependent limiting factors, 278–279
Density-independent limiting factors, 279
Deoxyribonucleic acid (DNA) structure, theory of, 252
Deoxyribonucleic acid (DNA), 184–185, 315
Deoxyribose, 252
Dependent variable, 70, 119
Descriptive thought, 60–63
Differentiation, 198–200
Diffusion, 223
Disproof, 36–38, 67
Dominant factor, 160

Ecological niche, 282–283
Ecosystem dynamics, theory of, 284
Ecosystem, 142, 281
Ectoderm, 200
Egg, 159, 191, 195, 197
Electron transport chain, 239–240
Electrons, 141, 210, 212
Embryological development, theory of, 197–200
Emergent properties, 141–144
Endoderm, 200
Endothermic reaction, 219
Energy, 210
Environment, 273
Enzymes, 218, 222
Epigenesis, 197
Equatorial plate, 188–189
Eukaryote, 153
Evidence, 35–37, 47, 83
Evolution, synthetic theory of, 318
Evolution, Darwin's theory of, 309
Evolution, 153–154
Excretion, 282
Exothermic reaction, 219
Experiment, 35, 83
Experimental group, 68
Explanation, 42, 83
Exponential growth, 272, 276
Extinction, 308

Fact, 45
Fair test, 116–117
Families, 153
Fermentation, 236–237
Fertilization, 191
Food chain, 282–283
Food web, 282–283
Fossil record, 311, 323
Fungi, 153

Gametes, 158, 191
Gases, 211

Gastrula, 200
Gene pool, 318
Genera, 153
Genes, 160, 162, 184, 195, 218, 252
Genetic code, 252, 263
Genetic diversity, 193
Genetic potential, 157
Genotype, 163
Geological record, 314
Glucose, 217, 237
Glycolysis, 237
Gold, 211
Government, theory of, 46
Guttation, 177

Habitat, 273
Heart function, 173
Heart rate regulation, theory of, 174–176
Helium, 210
Herbivore, 281
Homeostasis, 174
Hydrogen, 210
Hydrogen ions, 248
Hydroxyl ions, 248
Hypothesis, 34, 47, 64, 83
Hypothetical thought, 60–63
Hypothetico-deductive thinking, 36–37, 43, 62, 65, 71, 83

Independent variable, 66, 70, 119
Independent assortment, 159, 162
Induction, 42
Inductive-deductive thinking, 38–41
Inheritance of acquired characteristics, theory of, 321
Inorganic molecules, 217, 281, 306
Instinct, 157
Insulin, 219
Interspecific competition, 277
Intraspecific competition, 277
Invagination, 200
Ion, 174
Iron, 211
Isotope, 211, 251, 253

Kinetic energy, 210, 237
Kinetic-molecular theory, 213
Kingdoms, 153
Knowledge acquisition, theory of, 47
Kreb's cycle, 239–240

Learning, 41, 157
Learning how-to-learn, 71
Levels of organization, 7, 141–143
Life, 140–141
Light energy, 281
Light, 245
Limiting factors, 272, 276, 313
Lipids, 217
Liquids, 211
Lithium, 211

Macro-to-micro approach, 7–11
Meiosis, 191–195
Mendel's theory of inheritance, 44, 162
Mendel's factors, 158–162, 185, 195
Mesoderm, 200
Metabolic rate, 219
Metabolism, 219
Metabolism, general theory of, 222
Micro-to macro approach, 7–11
Migration, 318
Mitochondria, 241
Mitosis, 187
Mnemonic devices, 22
Molecular movement, theory of, 226
Molecules, 141, 211
Multicellular organisms, 142
Mutations, 163, 315

NADP 248
NADPH, 248
Natural selection, Darwin's theory of, 45, 315
Neutron, 141, 210
Nucleic acid, 218
Nucleotide, 218, 268

Omnivore, 282
Optimum range, 274
Orders, 153
Organ system, 142
Organ, 142
Organelles, 141, 243
Organic molecules, 217, 308
Origin of life, theory of, 307
Osmosis, 176–177, 224

pH, 220–221
Phenotype, 162–163
Phloem, 179
Phosphate group, 218
Phosphoric acid, 252–253
Photons, 153, 210, 226, 244
Photosynthesis, 176, 244, 281
Photosynthesis, light-dependent reactions, 246–249
Photosynthesis, light-independent reactions, 249
Photosynthesis, theory of, 250
Phyla, 153
Phyletic evolution, 318
Plants, 153
Polarity, 177
Pollen, 158, 191
Population, 85, 142, 272
Population crash, 273
Population growth, a theory of, 274
Postulates, 44, 47
Potential energy, 212
Predator-prey interaction, 276
Prediction, 24, 34, 47, 65, 83
Preformation, 197
Pressure, 212
Primary succession, 286
Primitive atmosphere, 306
Probabilistic thinking, 62, 67–70, 81–88
Probability, 86, 107–108
Producers, 281
Prokaryote, 153
Proof, 36–38, 66–67
Proportion, inverse, 129
Proportion, direct, 127
Proportional thinking, 62
Proportions, 124
Protein synthesis, theory of gene function, 256–257
Proteins, globular, 218
Proteins, 217
Protist, 153
Protons, 141, 210, 248
Pure science, 43
Pyruvic acid, 237

Radioactive, 258
Reasoning, 157
Recessive factor, 160
Reciprocal relationship, 130
Reflex, 157
Regeneration, 186–187
Ribonucleic acid (RNA), 257–259
Root pressure, 176

S-shaped growth, 272–273
Sample, 86
Scavengers, 282
Science, 34
Scientific naming, 153
Secondary succession, 286
Selectively permeable, 223
Self-pollination, 158
Serial ordering, 60
Sexual reproduction, theory of, 194
Solids, 211

Special creation, theory of, 154, 310
Speciation, 318
Species, 153–156
Spectrophotometer, 251
Sperm, 158, 191, 195, 197
Spindle fibers, 190
Spontaneous generation, 137, 309
Subatomic particles, 141
Succession, theory of, 289
Suction, 177
Survival of the fittest, 314

Taxis, 157
Teleological, 29
Temperature, 212
Theory, 43–47
Thylakoids, 246
Tissues, 142
Traits, 152
Transpiration, 177, 285
Trophic levels, 281–282

Uranium, 211

Values, 66
Variable, 66, 89
Variation, 313–314
Vein, 172
Viruses, 140
Vitalism, 136

Water rise in plants, theory of, 176–178
Wave length, 244

Xylem, 176

Zygote, 161, 191, 198, 200